We Can Do It!

Year 5

USING AND APPLYING MATHS CHALLENGES
Peter Clarke

Acknowledgement

The author wishes to thank Mike Askew, Sheila Ebbutt and Brian Molyneaux for their valuable contributions to this publication.

Thanks also to the following:

Karen Holman, Paddox Primary School, Rugby
Hilary Head, Send C of E First School, Surrey
Shirley Mulroy, Leslie Rankin, Hady Primary School, Chesterfield
Catherine Aket, Whitecrest Primary School, West Midlands
Kim Varden, Broke Hall Primary School, Suffolk
Elaine Richardson, St Augustine's Primary School, Cheshire
Carolyn Wallis, St Nicholas House Junior School, Hertfordshire
Lyn Wickham, Emma Bailey, Bidbury Junior School, Hampshire
Mandy Patterson, Temple Mill Primary School, Kent
John Ellard, Kingsley Primary School, Northampton
Father Rudolf Loewenstein, St Christina's Primary School, Camden
Joyce Lydford, Balgowan Primary School, Kent
Louise Guthrie, Angela Beall, Bardsey Primary School, Leeds
Sharon Thomas, Cwmbwrla Primary School, Swansea
Lynwen Barnsley, Education Effectiveness, Swansea
Jennie Jump, Advisor, Leeds
Sharon Sutton, University of Reading
Steve Lumb, Fielding Primary School, Ealing
Deborah de Gray, West Kingsdown C of E Primary School, Kent
Kerry Ann Darlington, Ullapool Primary School, Ross-shire
Helen Elis Jones, University of Wales, Bangor
Jayne Featherstone, Elton Community Primary School, Lancashire
Jane Airey, Frith Manor Primary School, Barnet
Andrea Trigg, Felbridge Primary School, West Sussex
Helen Andrews, Blue Coat School, Birmingham
Joyce Atkinson, Croham Hurst Junior School, Surrey
Jane Holmes, Elizabeth Wyles, St John's Primary School, Oxon

Thanks also to the BEAM Development Group:

Mich Bahn, Canonbury Primary School, Islington
Joanne Barrett, Rotherfield Primary School, Islington
Catherine Horton, St Jude and St Paul's School, Islington
Simone de Juan, Prior Weston Primary School, Islington

Published by BEAM Education
Maze Workshops
72a Southgate Road
London N1 3JT

Telephone 020 7684 3323
Fax 020 7684 3334
Email info@beam.co.uk
www.beam.co.uk

© Peter Clarke and BEAM Education 2008, a division of Nelson Thornes

ISBN 978 1 9062 2449 3
British Library Cataloguing-in-Publication Data
Data available
Edited by Marion Dill
Design by Reena Kataria
Layout by Matt Carr
Illustrations by Matt Carr
Cover photo: Lauriston School, Hackney
Printed by Graphy Cems, Spain

Contents

Introduction

What is AT1: Using and applying mathematics?

Mathematical problem solving involves using previously acquired mathematical understanding, knowledge and skills and applying these to solve problems arising within everyday life, as well as within mathematics.

A major reason for studying mathematics is to be able to develop problem-solving, reasoning and logical skills that we can apply to everyday situations. We want children to employ their 'pure' mathematical knowledge effectively in real-life situations day by day, both within and outside school. It is important for children to see how acquiring mathematical understanding can help them solve problems that are relevant to their daily lives. Mathematics can help them interpret and analyse real-life situations. It can also be a source of creative pleasure for its own sake.

Using and applying mathematics is often mistaken as simply solving word problems.

Tomas shared 20 marbles equally among himself and his four friends. How many marbles did each child get?

To solve this word problem, you need to be able to read and understand what the problem is, identify the calculation you need to do, do the calculation and then interpret the answer you get in the context of the problem. In a limited sense, you are using and applying mathematics, but at a low level. Word problems at this level follow a predictable pattern, which removes the need for any real problem solving.

Of course, word problems can be more complex, and children have to work out which information is relevant, and what the context tells you about the kind of answer you need (and knowing key words is not always a helpful strategy). This multi-step problem is more like the kinds of problems that children face in real life.

Tomas is 8 years old, and he gets 24 marbles. He keeps half of the marbles for himself and shares out the rest equally among his four friends. How many more marbles does he have than each of his friends?

The point about more complex problems is that you have to work out what the meaning is, what sort of outcome you need, and what sensible calculations to do to get there. Problems in real life are mainly like this.

An investigative approach to the simple marble problem could be:

How many different ways can you share 20 marbles equally?

A more complex investigation could be:

It's Tomas's birthday, so he gets more marbles than anyone else. Everyone else gets the same amount. How many different ways can you share the 24 marbles?

Sometimes it is interesting just exploring some mathematics for its own sake, as a pastime. Some of the problems in **We Can Do It!** are like that. The interest lies in using your brain, finding a pattern, seeking a neat solution – just like doing a crossword or sudoku puzzle. In the process of working on the investigation, children will be honing their reasoning skills and using their creativity to seek a way forward.

Some problems really do mirror everyday life:

If we spend less time tidying up, will we get more time for playing outside?

What news stories on the front page of the newspaper are given the most space?

What proportion of water do you need to dilute a fruit drink?

Many of these involve measures and data handling. Children need practical experiences to solve these: if you don't know what 100 ml of liquid looks like, and how it compares with 1 litre, you cannot solve the problem in the abstract. This means being prepared for a busy, active and, at times, messy classroom – and also a classroom where children discuss with each other the problem in hand.

Real and complex problems and investigations require children to search for strategies to get started and to draw upon their experiences and knowledge of 'pure' mathematics. They also encourage children to work flexibly, creatively and logically. They are less comfortable for the teacher because the outcomes are not always predictable and the answers are not always known. Our role is to work with children, sometimes doing the mathematics alongside them, looking for and encouraging creative and logical thinking, rather than focusing on right answers.

Mathematical thinking

The *National Curriculum* (2000) outlines the thinking skills that complement the key understanding, knowledge and skills that are embedded in the statutory primary curriculum.

The *We Can Do It!* series aims to develop the following key thinking skills in children:

Information – processing skills

- Locate, collect relevant information
- Sort, classify, sequence, compare and analyse part and/or whole relationships

Reasoning skills

- Give reasons for opinions and actions
- Draw inferences and make deductions
- Use precise language to explain what they think
- Make judgements and decisions informed by reason or evidence

Enquiry skills

- Ask relevant questions
- Pose and define problems
- Plan what to do and how to research
- Predict outcomes and anticipate conclusions
- Test conclusions and improve ideas

Creative thinking skills

- Generate and extend ideas
- Suggest hypotheses
- Apply imagination
- Look for alternative innovative outcomes

Evaluative skills

- Evaluate information
- Judge the value of what they read, hear or do
- Develop criteria for judging the value of their own and others' work or ideas
- Have confidence in their judgement

We Can Do It!

In this series, we provide problems and challenges that stimulate genuine mathematical thinking. These problems are written for a community of learners in the primary classroom – that is, we expect the problems to be solved collaboratively by pairs of children, groups and whole classes working together and discussing the problems at every stage. With each problem, we offer teaching advice on how to encourage high-level thinking among children. We also analyse each problem and children's possible responses to it in order to promote greater understanding of how children develop problem-solving skills.

The challenges in *We Can Do It!* are designed to improve children's attainment in the three strands of AT1 of the *National Curriculum* (2000): Using and applying mathematics.

In **problem solving** by:

- using a range of problem solving strategies

- trying different approaches to a problem

- applying mathematics in a new context

- checking their results

In **communicating** by:

- interpreting information

- recording information systematically

- using mathematical language, symbols, notation and diagrams correctly and precisely

- presenting and interpreting methods, solutions and conclusions in the context of the problem

In **reasoning** by:

- giving clear explanations of their methods and reasoning

- investigating and making general statements

- recognising patterns in their results

- making use of a wider range of evidence to justify results through logical reasoned argument

- drawing their own conclusions

The challenges also provide children with an opportunity to practise and consolidate the five themes and objectives of Strand 1: Using and applying mathematics for Year 5 in the *Renewed Framework of Mathematics* (2006).

Creating a problem-solving classroom

It is important that children have faith in their own abilities and develop a healthy self-esteem. They need to be encouraged to have a go, even if at first their attempts are wrong. We want children to realise that having a go and making a mistake is far better than not attempting a problem at all, and that trial and improvement is a vital part of the learning process. Therefore it is important to encourage and reward the following qualities during problem-solving lessons:

- perseverance

- flexibility

- originality

- active involvement

- independence

- cooperation

- willingness to communicate and share ideas

- willingness to try and take risks

- reflection

Teacher expectations are a critical factor affecting children's achievement. We can engender a classroom ethos that makes anything possible for all children. We can offer children opportunities to reach their full potential, regardless of supposed appropriate year-level expectations.

Assessment

You can use the challenges in *We Can Do It!* with the whole class or with groups of children as an assessment activity. Linked to the strand that is being studied at present, *We Can Do It!* will not only provide you with an indication of how well the children have understood the 'pure' mathematics objectives being covered, but also their problem-solving skills.

Throughout each of the challenges there are prompting questions which focus on specific aspects of the challenge. At the end of each challenge there are also three questions that are specifically designed to help with assessing using and applying mathematics.

The list of thinking-skills statements on page 7 and the descriptions relating to the three strands of AT1 on page 8 are extremely useful in helping assess children's problem-solving skills.

Problem-solving strategies

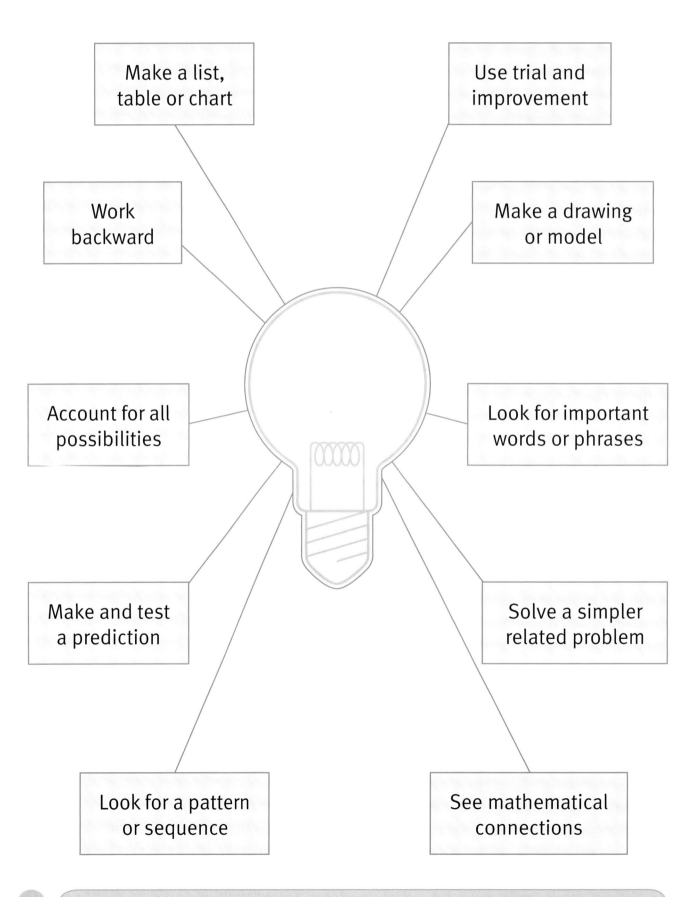

Make a list, table or chart

Use trial and improvement

Work backward

Make a drawing or model

Account for all possibilities

Look for important words or phrases

Make and test a prediction

Solve a simpler related problem

Look for a pattern or sequence

See mathematical connections

A final word

From an early age, children can learn that school mathematics is 'work' – a series of tasks they need to get through as quickly as possible, preferably without the need for thought. The challenges in **We Can Do It!** are deliberately demanding for children in order to promote their ability to solve problems. You will need to encourage them to rely less on your help, setting them off to work on a challenge for a short length of time. Follow this with time together to discuss the different ways in which they have set about the task; this will help children realise that they can achieve something, while you feed in ideas for continuing without taking the responsibility for the thinking away from the children.

Being challenged is enjoyable! The challenges in **We Can Do It!** have not been 'dressed up' to disguise the mathematics or to make them 'fun'. The aim is not to make mathematics itself enjoyable but rather find enjoyment by being prepared to have a go at something, rising to the challenge and reaching a satisfactory conclusion.

How to use this book

Question to pose the challenge

Opening question to ask the children that is designed to act as a springboard into the challenge

Summary of maths content

Brief summary of the 'pure' mathematics focus of the challenge

Introducing the challenge

Outline scenario to hook in the children's interest. It often includes opportunities to engage the children's interest further by including 'turn and talk' instructions.

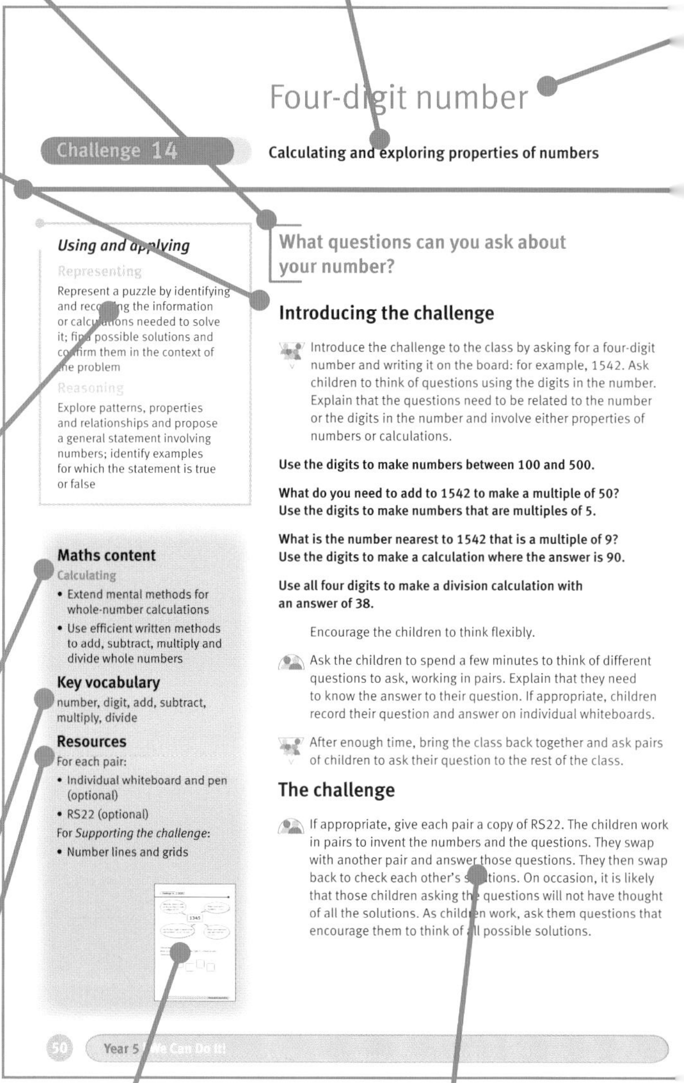

Using and applying

Description of how the challenge links to the five themes in Strand 1: Using and applying mathematics), in the *Renewed Framework for Mathematics* (2006)

Solving problems

Representing

Enquiring

Reasoning

Communicating

Maths content

Objectives from the *Framework* specifically covered in the challenge

Key vocabulary

List of words and phrases appropriate to the challenge

Resources

List of resources children need to undertake the challenge, including resource sheets (RS)

RS diagram

You find the resource sheet (RS) on the CD.

The challenge

This offers advice on how to structure the challenge and uses the following symbols for clarification:

 individual   paired

 group   whole class

Sample page content

Four-digit number

Challenge 14 — Calculating and exploring properties of numbers

Using and applying

Representing

Represent a puzzle by identifying and recording the information or calculations needed to solve it; find possible solutions and confirm them in the context of the problem

Reasoning

Explore patterns, properties and relationships and propose a general statement involving numbers; identify examples for which the statement is true or false

Maths content

Calculating

- Extend mental methods for whole-number calculations
- Use efficient written methods to add, subtract, multiply and divide whole numbers

Key vocabulary

number, digit, add, subtract, multiply, divide

Resources

For each pair:

- Individual whiteboard and pen (optional)
- RS22 (optional)

For *Supporting the challenge*:

- Number lines and grids

What questions can you ask about your number?

Introducing the challenge

Introduce the challenge to the class by asking for a four-digit number and writing it on the board: for example, 1542. Ask children to think of questions using the digits in the number. Explain that the questions need to be related to the number or the digits in the number and involve either properties of numbers or calculations.

Use the digits to make numbers between 100 and 500.

What do you need to add to 1542 to make a multiple of 50?
Use the digits to make numbers that are multiples of 5.

What is the number nearest to 1542 that is a multiple of 9?
Use the digits to make a calculation where the answer is 90.

Use all four digits to make a division calculation with an answer of 38.

Encourage the children to think flexibly.

Ask the children to spend a few minutes to think of different questions to ask, working in pairs. Explain that they need to know the answer to their question. If appropriate, children record their question and answer on individual whiteboards.

After enough time, bring the class back together and ask pairs of children to ask their question to the rest of the class.

The challenge

If appropriate, give each pair a copy of RS22. The children work in pairs to invent the numbers and the questions. They swap with another pair and answer those questions. They then swap back to check each other's solutions. On occasion, it is likely that those children asking the questions will not have thought of all the solutions. As children work, ask them questions that encourage them to think of all possible solutions.

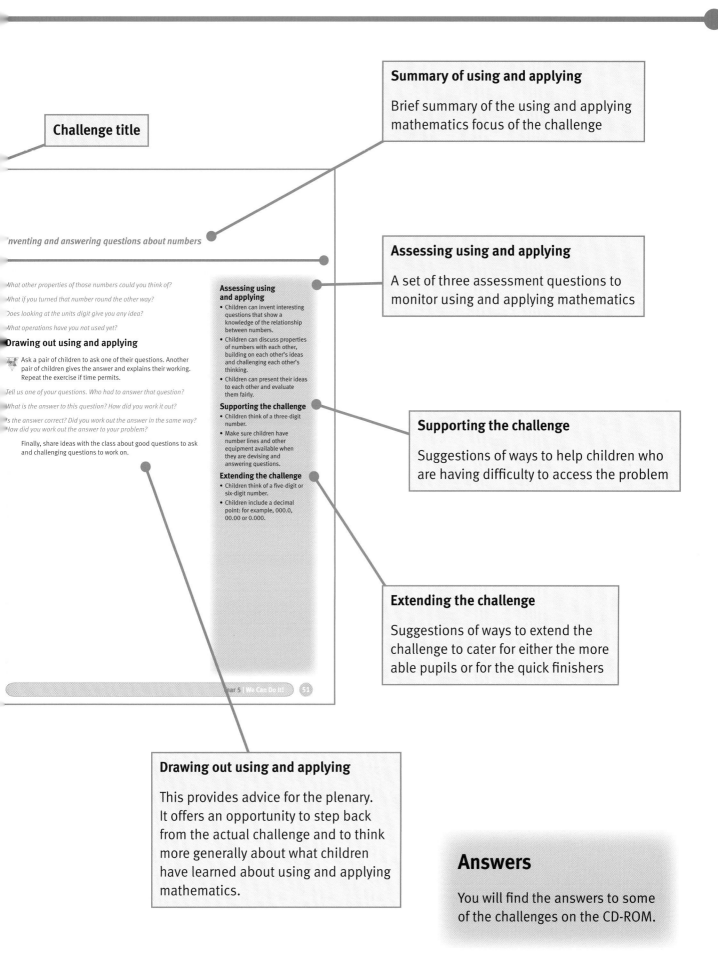

Challenge title

Summary of using and applying

Brief summary of the using and applying mathematics focus of the challenge

Assessing using and applying

A set of three assessment questions to monitor using and applying mathematics

Supporting the challenge

Suggestions of ways to help children who are having difficulty to access the problem

Extending the challenge

Suggestions of ways to extend the challenge to cater for either the more able pupils or for the quick finishers

Drawing out using and applying

This provides advice for the plenary. It offers an opportunity to step back from the actual challenge and to think more generally about what children have learned about using and applying mathematics.

Answers

You will find the answers to some of the challenges on the CD-ROM.

Inventing and answering questions about numbers

What other properties of those numbers could you think of?

What if you turned that number round the other way?

Does looking at the units digit give you any idea?

What operations have you not used yet?

Drawing out using and applying

Ask a pair of children to ask one of their questions. Another pair of children gives the answer and explains their working. Repeat the exercise if time permits.

Tell us one of your questions. Who had to answer that question?

What is the answer to this question? How did you work it out?

Is the answer correct? Did you work out the answer in the same way? How did you work out the answer to your problem?

Finally, share ideas with the class about good questions to ask and challenging questions to work on.

Assessing using and applying
- Children can invent interesting questions that show a knowledge of the relationship between numbers.
- Children can discuss properties of numbers with each other, building on each other's ideas and challenging each other's thinking.
- Children can present their ideas to each other and evaluate them fairly.

Supporting the challenge
- Children think of a three-digit number.
- Make sure children have number lines and other equipment available when they are devising and answering questions.

Extending the challenge
- Children think of a five-digit or six-digit number.
- Children include a decimal point: for example, 000.0, 00.00 or 0.000.

Lesson suggestions

Aspects of mathematical problem solving should be covered in every maths lesson, even those that aim to teach the purest of mathematical concepts. Children need to see the application of 'pure' maths in everyday experiences.

We advise, however, that once a week you devote a lesson entirely to developing children's problem-solving skills. It is for this reason that *We Can Do It!* consists of 36 challenges.

The challenges in this book provide children with an opportunity to practise and consolidate the Year 5 objectives from the *Renewed Framework for Mathematics* (2006). The curriculum charts on pages 18-21 show which challenge is matched to which planning block and mathematics strand. Refer to these charts when choosing a challenge.

We Can Do It! and the daily maths lesson

The challenges in *We Can Do It!* are ideally suited for the daily maths lesson. You can introduce each challenge to the whole class or to groups of children. Here is a suggestion how to structure a lesson using *We Can Do It!*.

Introducing the challenge

- Introduce the idea of the challenge either as a discussion or by giving a simplified version of the problem. You may need to highlight some of the mathematics children need to solve the problem.

- Introduce the challenge to the children by asking the question that poses the problem.

- Stimulate children's involvement through discussion.

- Use the key vocabulary throughout and explain new words where necessary.

- Make sure that the children understand the challenge.

- If you use a resource sheet, make sure children understand the text on the sheet.

- Begin to work through the challenge with the whole class, pointing out possible problem-solving strategies.

The challenge

- Arrange children into pairs or groups to work on the problem.

- Make sure appropriate resources are available to help children with the challenge.

- Monitor individuals, pairs or groups of children, offering support when and where needed.

- If appropriate, extend the challenge for some children.

Drawing out using and applying

- Plan an extended plenary.

- Discuss the challenge with the class.

- Invite individual children, pairs or groups to offer their solutions and the strategies they used.

The teacher's role in problem-solving lessons

- Give a choice where possible.

- Present the problem orally, giving maximum visual support where appropriate.

- Help children 'own the problem' by linking it to their everyday experiences.

- Encourage children to work together, sharing ideas for tackling a problem.

- Allow time and space for collaboration and consultation.

- Intervene, when asked, in such a way as to develop children's autonomy and independence.

- Work alongside children, setting an example yourself.

- Encourage the children to present their work to others.

Paired and group work

We Can Do It! recognises the importance of encouraging children to work collaboratively. All of the challenges in *We Can Do It!* include some element of paired or group work. By working as a group, children develop cooperation and collective responsibility. They also learn from each other, confirming their mathematical knowledge and identifying for themselves, in a non-threatening environment, any misconceptions they may hold.

The *National Curriculum* identifies three strands of the AT1: Using and applying mathematics. They are problem solving, communicating and reasoning. While it is possible for children to problem solve independently, communication, as the diagram below illustrates, is a cooperative, interactive process that involves both expressing and receiving information.

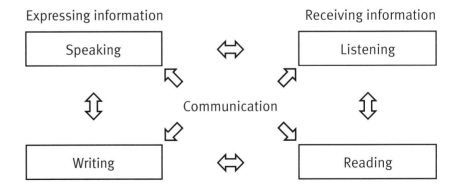

Meaningful reasoning can only occur through communication. Children cannot effectively reason with themselves: they always see themselves as being right! It is not until they begin to discuss and share ideas with others that children begin to reason and see other points of views and possibilities.

Charts linking to the
Renewed Framework for Mathematics (2006)

Chart linking blocks and strands of the *Renewed Framework for Mathematics* (2006)

	Strand 2: Counting and understanding number	Strand 3: Knowing and using number facts	Strand 4: Calculating	Strand 5: Understanding shape	Strand 6: Measuring	Strand 7: Handling data
BLOCK A: Counting, partitioning and calculating	●	●	●			
BLOCK B: Securing number facts, understanding shapes		●		●		
BLOCK C: Handling data and measures					●	●
BLOCK D: Calculating, measuring and understanding shape			●	●	●	
BLOCK E: Securing number facts, calculating, identifying relationships	●	●	●			

Chart linking challenges in *We Can Do It! Year 5* to strands of the *Renewed Framework for Mathematics* (2006)

Challenge Number	Title	Strand 1: Using and applying mathematics					Strand 2: Counting and understanding number	Strand 3: Knowing and using number facts	Strand 4: Calculating	Strand 5: Understanding shape	Strand 6: Measuring	Strand 7: Handling data
		Solving problems	Representing	Enquiring	Reasoning	Communicating						
1	How many zeros?		●		●		●					
2	Decimal remainders		●		●		●					
3	1–100 percentages	●	●				●					
4	Percentage squares				●	●	●		●			
5	Fraction sort		●			●	●					
6	Feeding groups			●		●	●					
7	Money in my purse	●			●		●	●				
8	Number strings		●		●			●	●			
9	Domino puzzle	●			●			●				
10	Almost magic				●	●		●				

Number	Title	Strand 1: Using and applying mathematics					Strand 2: Counting and understanding number	Strand 3: Knowing and using number facts	Strand 4: Calculating	Strand 5: Understanding shape	Strand 6: Measuring	Strand 7: Handling data
		Solving problems	Representing	Enquiring	Reasoning	Communicating						
11	Factors				●	●		●				
12	Common multiples				●	●		●				
13	In a year			●		●		●	●			
14	Four-digit number		●		●				●			
15	Float for the stall	●				●			●			
16	How much do you eat?	●		●				●	●			
17	Heartbeat			●		●		●	●			
18	Sports gear	●		●					●			
19	Three in a line		●		●				●			
20	Magic 999		●		●			●	●			
21	Fraction cards	●	●						●			
22	Find the missing alien		●		●					●		
23	Front and sides		●		●					●		

Number	Title	Strand 1: Using and applying mathematics					Strand 2: Counting and understanding number	Strand 3: Knowing and using number facts	Strand 4: Calculating	Strand 5: Understanding shape	Strand 6: Measuring	Strand 7: Handling data
		Solving problems	Representing	Enquiring	Reasoning	Communicating						
24	Squares in squares				●	●				●		
25	Cross-stitch patterns				●	●				●		
26	Nine squares		●		●					●	●	
27	Egyptian circles			●		●				●	●	
28	Pave it	●			●					●	●	
29	Cutting a strip	●			●						●	
30	How strong?		●	●		●					●	
31	Hands and feet				●	●					●	
32	Round the world			●		●			●		●	
33	A fair six?				●	●						●
34	How likely?				●	●						●
35	Words in a newspaper		●	●								●
36	Channel hopping			●		●						●

The challenges

Strand 2: Counting and understanding number

Strand 3: Knowing and using number facts

Strand 4: Calculating

Strand 5: Understanding shape

Strand 6: Measuring

Strand 7: Handling data

How many zeros? ✓

Reading, writing and partitioning whole numbers to 1000

Using and applying

Representing

Represent a problem by identifying and recording the information needed to solve it; find possible solutions and confirm them in the context of the problem

Reasoning

Explore patterns, properties and relationships, and propose a general statement involving numbers

Maths content

Counting and understanding number

- Count from any given number in whole-number steps
- Explain what each digit represents in whole numbers, read and write these numbers

Key vocabulary

number, digit, figure, zero

Resources

- RS1 (optional, for each child)
- 100-grids (optional)

How many plastic zeros do you need for all the room numbers from 1 to 1000?

Introducing the challenge

 Tell the children that you met someone who was working on a grand new hotel that was being built. You met this person while out shopping and they were having to buy plastic numerals to put on the doors of the bedrooms. There were 1000 rooms in the hotel, and the person was wondering if they were going to need to buy the same number of each numeral. Did they need the same number of ones and zeros? Ask the children to turn and talk about what they think.

After you have discussed their various responses, suggest that one way to think about this problem is to try something simpler: for example, just the numbers from 0–20. Ask the children to imagine writing all the numbers from zero to 20.

How many zeros would you have to write? (3)

Write these on the board as a list. For example:

> 0   10   20

 Ask the children to imagine all the numbers from zero to 99 written down.

How many zeros would you have to write now? (10)

Can you explain why there are ten?

What about from zero to 100?

Continue the list on the board:

0   10   20   30   40   50   60   70   80   90   100

The challenge

 Introduce the challenge to the children, using RS1. Ask the children to think carefully before they start about how best to organise their working. It may be helpful to stop the children after a few minutes and ask them to share with each other the methods they are using to keep track of the number of zeros.

Will it help you to know how many zeros there are from 1 to 50 or from 1 to 100?

Will this help you work out how many zeros there are for the next 50 or 100 numbers? How?

What about from 1 to 1000?

Drawing out using and applying

Invite pairs to share their results with the class.

Did anyone get a different number of zeros?

How many plastic zeros will the hotel need for rooms 200 to 300? Rooms 151 to 251? Rooms 235 to 534?

How would you work that out?

Children explain the different methods they used and also the way they recorded their results.

How did you keep a track of how many zeros there were?

Which method do you think was the most effective? Why?

How did you break it down into manageable bits? What helped most?

What if you were going to count all the zeros in the numbers to one million? Which method would work best?

Assessing using and applying

- Children can write down all the numbers from 1 to 1000 with zero in them in order to count the zeros.
- Children can group the numbers to make counting the zeros more effective.
- Children can devise a system for keeping track of the zeros that would allow them to extend the problem to 1 000 000.

Supporting the challenge

- Children use a 100-grid and mark the zeros.
- Help the children set out the numbers in a systematic way so they can easily identify the pattern. For example, they can list the numbers from 1 to 100 in tens.

Extending the challenge

- How many zeros do you need to create all the numbers from 1 to 1 000 000?
- How many separate digits would you need to create all the numbers from 1 to 100? What about from 1 to 1000?

Decimal remainders

Relating fractions to division and their decimal equivalents

Using and applying

Representing

Represent a problem by identifying and recording the information or calculations needed to solve it; find possible solutions and confirm them in the context of the problem

Reasoning

Explore patterns, properties and relationships, and propose a general statement involving numbers; identify examples for which the statement is true or false

Maths content

Counting and understanding number

- Explain what each digit represents in decimals with up to two places and partition, round and order these numbers

- Express a smaller whole number as a fraction of a larger one: for example, recognise that 5 out of 8 is five eighths; relate fractions to their decimal representations

Key vocabulary

fraction, decimal, equivalent

Resources

- RS2, RS3
- RS4 (optional, for each group)
- Pencil, paper, calculator (for each child)

> ## Which fractions give a decimal answer to exactly two decimal places?

Introducing the challenge

 Randomly distribute the cards from RS2 and RS3 to the children.

Invite all the children who have the equivalent of a half to stand ($1 \div 2$, $\frac{1}{2}$ and 0.5). Collect these cards and the children sit down.

Write $3 \div 10$ on the board. Ask the children to tell you this as a fraction ($\frac{3}{10}$) and as a decimal (0.3). Repeat the above, writing $\frac{2}{5}$, then 0.2, asking the children to say the equivalent division calculation, fraction and decimal.

Next, ask a child to stand and show their card to the rest of the class: for example, $\frac{3}{5}$. Ask the rest of the class to look at their cards and the two children who have the equivalent division calculation ($3 \div 5$) and decimal (0.6) to stand also. Collect these three cards and ask another child to stand and show the rest of the class their card. Continue until the children have matched and collected all the cards.

Division	Fraction	Decimal
$1 \div 2$	$\frac{1}{2}$	0.5
$2 \div 5$	$\frac{2}{5}$	0.4
$3 \div 5$	$\frac{3}{5}$	0.6
$4 \div 5$	$\frac{4}{5}$	0.8
$1 \div 4$	$\frac{1}{4}$	0.25

Division	Fraction	Decimal
$3 \div 4$	$\frac{3}{4}$	0.75
$1 \div 5$	$\frac{1}{5}$	0.2
$1 \div 3$	$\frac{1}{3}$	0.333
$3 \div 2$	$\frac{3}{2}$	1.5
$4 \div 2$	$\frac{4}{2}$	2

The challenge

 Arrange the children into groups of four and introduce the challenge to the children, using RS4 if appropriate.

 Children briefly discuss the challenge as a group and decide on an approach to solving the problem.

It is important that children work effectively as a group. Working individually, using trial-and-improvement techniques, can be disheartening. Groups need to coordinate their work on division of one-digit numbers by other one-digit numbers until they begin to see a pattern and can start to make a generalisation. Groups also need to develop an effective record-keeping system.

 After enough time, bring the class back together and ask individual groups to share their initial thoughts with the rest of the class. It is helpful at this stage to share useful strategies.

 Children continue with the challenge in their groups. Monitor each group and, where appropriate, ask questions to prompt their thinking.

What type of answer will you get if you divide a larger number by a smaller number?

What if you divide a smaller number by a larger number?

How are you keeping track of what numbers you have divided?

How did you find some numbers where the answer is a decimal to two decimal places?

What do you notice about these numbers?

Can you make a prediction about a pair of numbers that might result in two decimal places?

Drawing out using and applying

 Ask individual groups to report back to the rest of the class. Encourage the groups to talk about the following:

- how they arrived at the answers;

- the roles that different members of the group took;

- how they kept a record of their work;

- what generalisations they can make.

Assessing using and applying

- Children can work independently, choosing numbers at random to identify some decimals to two decimal places.

- Children can share the task between the group and identify all the decimals to two decimal places.

- Children can make careful judgements about which numbers to choose, find some decimals to two decimal places and use these to make a generalisation before identifying the remaining decimals.

Supporting the challenge

- Children use a calculator to identify the decimals.

- Where necessary, help groups share the challenge between themselves and organise their results in a systematic way.

Extending the challenge

- How many calculations have an answer to exactly one decimal place?

- What if you divide a two-digit number by a one-digit number? Can you make any predictions?

1–100 percentages

Expressing groups of numbers as percentages

Using and applying

Solving problems

Solve problems involving whole numbers

Representing

Represent a problem by identifying and recording the information needed to solve it; find possible solutions and confirm them in the context of the problem

Maths content

Counting and understanding number

- Understand percentage as the number of parts in every 100
- Express tenths and hundredths as percentages

Key vocabulary

per, per cent, percentage, in every, multiples, factors, odd, even

Resources

- NNS ITP: Number grid

For each child:

- RS5
- Pencil and paper

What percentage of the numbers on a 100-grid are ...?

Introducing the challenge

 Display the NNS ITP: Number grid and remind the children that each square represents 1%. Highlight the numbers 1 to 10 and ask the children to say what per cent of the grid you have shaded (10%).

1	2	3	4	5	6	7	8	9	10
11	12	13	14	15	16	17	18	19	20
21	22	23	24	25	26	27	28	29	30
31	32	33	34	35	36	37	38	39	40
41	42	43	44	45	46	47	48	49	50

Repeat the above, highlighting other sections of the grid and asking the children to say the percentage. Invite them to suggest examples of percentages for you to highlight.

6	7	8	9	10
16	17	18	19	20
26	27	28	29	30
36	37	38	39	40
46	47	48	49	50
56	57	58	59	60
66	67	68	69	70
76	77	78	79	80
86	87	88	89	90
96	97	98	99	100

1	2	3	4	5	6	7	8	9	10
11	12	13	14	15	16	17	18	19	20
21	22	23	24	25	26	27	28	29	30
31	32	33	34	35	36	37	38	39	40
41	42	43	44	45	46	47	48	49	50
51	52	53	54	55	56	57	58	59	60

If necessary, write up the '%' sign and remind the children that 'per' means 'in every' as in '60 miles per hour' or '£10 per hour' and that 'cent' refers to the number 100 as in 'centimetre' and 'century'.

What does the term 'per cent' mean?

What does 100% mean? (the whole amount)

The challenge

 Give each child a copy of RS5 and introduce the challenge to the children. The first part of the challenge asks children to identify what per cent of the numbers on a 100-grid have specific number properties.

The second part of the challenge requires children to use and apply their knowledge of the number system and to write questions for a given percentage. If children find it difficult to write questions for 12%, ask them to write questions involving a given tenth such as 30% or 80%.

It may be necessary to discuss different possibilities for questions: for example, questions involving place value (tenths or units), multiples, factors, 'more than' and 'less than', and 'in between'.

 For the final part of the challenge, children write five or six questions for a friend to solve. They write the answers on a separate piece of paper. Make sure that children swap their questions with a child of similar ability. When the children have answered their friend's questions, they swap back and check each other's answers, discussing any differences.

Drawing out using and applying

 Briefly discuss the answers for the first part of the challenge before moving on to asking the children to share some of their questions with the rest of the class. Occasionally, ask individual children to offer one of their questions for the rest of the class to answer. Make sure that you have identified already that these questions previously have a definite answer.

What percentage of the 100-grid consists of odd numbers? If we know that, then what per cent of numbers are even?

Read out a question that has 25% or 12% as the answer.

Who can tell us another question with an answer of 25% or 12%?

Who has got a tricky question they would like the rest of the class to solve?

Assessing using and applying

- Children can use the 100-grid to identify percentages.
- Children can identify percentages and can ask questions related to known percentages such as tenths.
- Children can answer and create their own questions for any percentage.

Supporting the challenge

- Encourage the children to write questions based on known percentages such as tenths.
- Make sure children swap their questions with a child of similar ability. They can work together on the challenge in pairs.

Extending the challenge

- Challenge children to write the trickiest question they can for the 100-grid. They also need to work out the answer.
- Children write questions that have percentages as answers based on their own experiences. For example: "What per cent of children sitting at the table nearest the door are wearing trainers?"; "Our morning break is 20 minutes. What per cent of an hour is that?"

Percentage squares

Calculating simple percentages and explaining equivalences

Using and applying

Reasoning

Explore patterns, properties and relationships, and propose a general statement involving numbers; identify examples for which the statement is true or false

Communicating

Explain reasoning

Maths content

Counting and understanding number

- Understand percentage as the number of parts in every 100 and express tenths and hundredths as percentages

Calculating

- Find percentages of numbers and quantities

Key vocabulary

per cent, percentage, fair

Resources

- Giant dice with stickers showing 0%, 5%, 10%, 15%, 20%, 25% (optional)

For each pair:

- RS6, paper clip and pencil (for the spinner)

For each child:

- Two copies of RS7
- Colour pencil

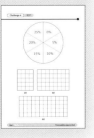

Are both these games fair?

Introducing the challenge

 Draw two grids of 20 squares on the board. Play the first game on RS6, you against the class (or the class in two teams). You could use the spinner on the sheet or roll a giant dice with stickers showing the percentages.

Invite children to share their strategies for calculating mentally various percentages.

If you know that 10% of 40 is 4, how can you work out what 5% of 40 is?

What about 20% of 40?

The challenge

 Children play both versions of the game. Encourage them to use strategies for calculating mentally the percentages, for example, saying:

> **I know 10% of 60 is 6 and 5% of 60 is 3, so 15% of 60 must be 9.**

Once the children have played both games, they discuss the fairness of both versions of the game.

Drawing out using and applying

 Ask the children to discuss whether or not they think both versions of the game are fair.

Can anyone explain why the size of the initial grid does not matter?

Do you think both versions of the game are fair?

If so, why? If not, why not?

Assessing using and applying

- Children can explain why Game 1 is fair, but cannot explain why Game 2 is fair.
- Children can decide that both games are fair on the basis of the outcome of rounds played.
- Children can explain why both games are fair.

Supporting the challenge

- Provide the children with a table showing 5%, 10%, 15%, 20% and 25% of 20, 40 and 60.

	20	40	60
5%	1	2	3
10%	2	4	6
15%	3	6	9
20%	4	8	12
25%	5	10	15

- After each spin of the spinner, encourage the child to say out loud the percentage calculation – for example, 25% of 40 is 10 – and for their partner to check their calculation, using the above table.

Extending the challenge

- Children try different-sized grids.
- Children play with a dice or spinner marked with fractions: for example, $\frac{1}{2}$, $\frac{3}{8}$, and so on. What would be a good size grid to play on?

Fraction sort

Recognising the relationship between fractions, decimals and percentages

Using and applying

Representing

Represent a problem by identifying and recording the information or calculations needed to solve it; find possible solutions and confirm them in the context of the problem

Communicating

Explain reasoning using diagrams and text; refine ways of recording using images and symbols

Maths content

Counting and understanding number

- Relate fractions to their decimal representations
- Understand percentage as the number of parts in every 100 and express tenths and hundredths as percentages

Key vocabulary

fraction, vulgar fraction, decimal fraction, percentage, symbol, diagram, story, calculation, representation, equivalent

Resources

For each pair:

- RS8, RS9
- Scissors

For *Supporting the challenge*:

- Squared paper

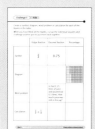 

> Can you create a symbol, diagram, word problem or calculation to complete the fractions, decimals and percentages table?

Introducing the challenge

 On the board, draw a diagram representing a fraction.

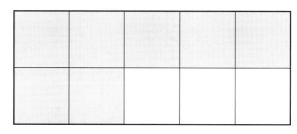

Invite a child to write the fraction symbol to represent the shaded part ($\frac{7}{10}$). Invite the children to suggest a calculation with this fraction as its answer.

What calculation might have $\frac{7}{10}$ as the answer? As part of the problem?

Children offer the decimal and percentage representation of the fraction.

What would you do to describe the shaded part as a decimal?

How would you describe it as a percentage?

What word problem might have $\frac{7}{10}$, 0.7 or 70% as the answer?

The challenge

 Provide each pair with a copy of RS8 and a pair of scissors. Introduce the challenge to the children. Encourage them to work together and agree on what to put in each cell on the grid before actually filling it in. You may need to make sure that the word problem matches the calculation they devise, not just that it has the same answer.

*What is three quarters as a percentage? What percentage is
the remaining bit?*

*Can you think of a diagram to show three quarters as a fraction?
Would this work as a diagram to show three quarters as a decimal,
or will you have to do a different one?*

What is the calculation in your word problem?

When the pairs have completed the table, they cut up the
table and swap with another pair who try and put the table
back together, using RS9. They then swap back to check
each other's completed table.

Drawing out using and applying

 Invite pairs of children to offer suggestions to complete
the table.

*How many different diagrams do we have for three quarters?
Do they all work? How do you know?*

*Read a word problem to the class. What is the calculation that
describes your word problem? Is there a different word problem
that describes the same calculation?*

What other calculation works for three quarters? Can you explain?

Finally, discuss the challenge with the children.

*Which missing part of the table only had one possible correct
answer?*

Which parts of the table had lots of different possible answers?

What was the hardest part of this challenge? Why?

What have you learnt from this challenge?

Assessing using and applying

- Children can connect some
  representations, but only a
  limited number.
- Children can create pictures
  and diagrams, but have
  difficulty putting the ideas
  into context.
- Children can coordinate
  the different forms of
  representation.

Supporting the challenge

- Change the fractions:
  for example, $\frac{1}{2}$, $\frac{1}{4}$ or $\frac{3}{10}$.
- Children use squared paper
  to represent the fractions,
  decimals and percentages.

Extending the challenge

- Change the fractions:
  for example, $\frac{3}{5}$, $\frac{2}{3}$ or $\frac{9}{20}$.
- Children make up their own,
  partially completed grids
  to swap with each other,
  using RS9.

Feeding groups

Solving a problem involving proportion

Using and applying

Enquiring

Plan and pursue an enquiry; present evidence by collecting, organising and interpreting information

Communicating

Explain reasoning using diagrams and text; refine ways of recording using images and symbols

Maths content

Counting and understanding number

- Use sequences to scale numbers up or down
- Solve problems involving proportions of quantities

Key vocabulary

quantities, scale, proportion

Resources

- RS10

For each pair:
- RS11
- Pencil and paper
- Calculator

For *Extending the challenge*:
- Children's recipe books

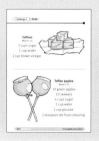

How much of each ingredient do you need to feed groups of different sizes?

Introducing the challenge

 Using RS10, display the two recipes to the class, either on the interactive whiteboard or as an enlarged poster. Set the scene for the children by explaining that a school is having a fete and that the sweet stall is going to sell, among other things, toffees and toffee apples. Referring to the recipes, discuss with the children how 24 toffees and 10 toffee apples would not be enough.

How many toffees and toffee apples do you think the school should make?

How many toffees does this recipe make? How many toffee apples?

How are we going to change the recipes so that there are enough ingredients to make that number?

Choose an appropriate number of toffees and toffee apples to make, making sure that the toffees are multiples of 24 and the toffee apples are multiples of 10.

Toffees (Recipe makes 24)	Toffee apples (Recipe makes 10)
48 (2 × 24)	50 (5 × 10)
72 (3 × 24)	80 (8 × 10)
120 (5 × 24)	100 (10 × 10)
240 (10 × 24)	200 (20 × 10)

For each recipe, work with the class to identify the quantities of each ingredient you need.

How many cups of water do you need to make 120 toffees? How much brown vinegar do you need?

How do we work that out?

Is there another way we could work it out?

Plan, collect and organise information
to answer a problem

The challenge

 Arrange the children into pairs. Provide each pair with a copy of RS11, pencil and paper and a calculator. Introduce the challenge to the children.

Encourage the children to think carefully about how they are going to present each recipe for different numbers of people. Will they write out each recipe separately for different groups of people or will they present the information in a table?

How much ... would you need to buy to feed 10 people?

What about the ...?

How did you work it out?

How are you going to write out the recipes?

Drawing out using and applying

 Invite individual pairs of children to suggest the quantities of different ingredients needed to feed different group sizes. Encourage children to explain how they worked out these quantities and how they recorded all the different quantities for the different ingredients.

How many onions do you need to feed 12 people? And 10 people?

How did you work it out?

Did anyone work it out a different way?

Is there another way you could have worked it out?

How did you keep a record of all the quantities needed for the different numbers of people?

Assessing using and applying

- Children can use doubling and halving to scale quantities up and down.
- Children can increase and decrease quantities to solve the problem.
- Children can use known quantities to work out unknown quantities: for example, combining the quantities needed for 4 people and 2 people when wanting to know the quantities to feed 6 people.

Supporting the challenge

- Children work with 2, 4 and 8 people and use doubling to work out the quantities.
- Help children organise their work in a systematic way: for example, in a table.

Ingredient	Number of people				
	2	4	6	10	12

Extending the challenge

- Children work out approximate costings for each of the items. They then work out the approximate total bill for groups of different sizes and the cost per person.
- Using children's recipe books, children plan their ideal menu for a party of 10 people. They then work out the quantities of each ingredient needed.

Money in my purse

Adding pounds and pence

Using and applying

Solving problems

Solve problems involving whole numbers

Reasoning

Explore patterns, properties and relationships

Maths content

Counting and understanding number

- Use knowledge of rounding, place value, number facts and inverse operations to estimate and check calculations

Knowing and using number facts

- Use knowledge of place value and addition and subtraction of two-digit numbers to derive sums and differences

Key vocabulary

add, total, sum, coin, pence, amount

Resources

- Demonstration coins: two 1p, two 2p and one 5p

For each pair:

- Pencil and paper
- RS12 (optional)

For *Supporting the challenge*:

- 1p, 2p, 5p, 10p and 20p coins

What different amounts could I make?

Introducing the challenge

 Display the five coins to the class.

Children work out the total amount of money on display (11p), then think of all the different amounts they could pay, using some or all of these coins, where they do not receive any change.

How much money do I have to spend?

What would be the price of the cheapest item I could pay for, using these coins?

What are the different amounts I could pay for, using some or all of these coins?

 Write up a few suggestions, then ask the children to work in pairs to identify all the different amounts possible.

 Bring the class back together and collect results. Write these on the board. Continue until you have written up all the amounts possible.

Briefly discuss with the class the different possibilities of making certain amounts of money: for example, 2p can be made from one 2p coin or using two 1p coins; 4p can be made using two 2p coins or a 2p coin and two 1p coins.

At this stage, do not ask the children to discuss their way of working and recording.

The challenge

 Briefly introduce the challenge to the children using RS12. Make sure that the children realise that they have to identify the different amounts possible, and also those amounts up to 38p that cannot be paid for exactly, and to explain why they think this is.

The children work in pairs, although they may prefer to do some of the working out individually and check progress with each other.

As the children work, look out for points for discussion and, where appropriate, draw the children together to sort errors and misconceptions before they continue.

How are you going to organise your work?

Can anyone show a list or table or a chart that would help?

How do you know you haven't missed any?

As the children progress with the activity to the last question, ask:

What is the first amount you cannot make? Why is that?

Drawing out using and applying

 Children present their results neatly in the form of a table for a class display, showing the structure of their working and thinking.

Can you show and explain to the rest of the class how you went about finding all the different amounts?

How did you keep a record of the different amounts you could pay for?

How did you make sure that you didn't miss out on any amounts?

Did any other pair record their working a different way? What did you do?

Which systems of recording were better? Why?

Finally, children write a short justification why they think certain amounts are missing. Some children can read these out for general comment and discussion.

Which amount could you not pay for using the 1p, 2p, 5p, 10p and 20p coin? Why is this?

Does anyone have a different explanation?

Assessing using and applying

- Children can work and record in such a way that shows they are thinking logically and systematically.
- Children can explain why they think they have found all the amounts possible.
- Children can give clear reasons why they cannot make certain amounts.

Supporting the challenge

- Children use coins to set out to work out all the different amounts possible.
- Children find different amounts using one coin, then two coins, three coins, four coins and all five coins.

Extending the challenge

- Add a 50p coin to the purse and encourage children to investigate the amounts they can make.
- Incorporate decimal notation into the challenge by adding a £1 coin to the purse.

Number strings

Using all four operations, including brackets; applying the order of operations

Using and applying

Representing

Represent a puzzle by identifying and recording the information or calculations needed to solve it; find possible solutions and confirm them in the context of the problem

Reasoning

Explore patterns, properties and relationships and propose a general statement involving numbers; identify examples for which the statement is true or false

Maths content

Knowing and using number facts

- Use knowledge of place value and addition and subtraction/ multiplication and division

Calculating

- Extend mental methods for whole-number calculations

Key vocabulary

add, subtract, multiply, divide, brackets

Resources

For each pair:

- RS13 (optional), pencil and paper

For *Supporting the challenge*:

- Number lines and grids

What is the longest string of consecutive numbers you can make?

Introducing the challenge

 Ask the children to suggest four different one-digit numbers and write these on the board. For example:

1, 3, 7, 8

Show how you can combine these four digits in any way, using the four operations and brackets to create different totals. For example:

$$12 = 83 - 71 \qquad 2 = 7 + 8 - 13 \qquad 19 = 8 + 7 + 3 + 1$$

Ask the children to offer suggestions and work with them on the use of brackets to make their intentions unambiguous. For example:

$$3 \times 8 - 7 + 1 \text{ could be } 18 = ((3 \times 8) - 7) + 1$$

or

$$4 = (3 \times (8 - 7)) + 1$$

The challenge

 Introduce the challenge to the children using RS13. For the first 10 minutes of the activity, all the pairs work with the same four digits given on RS13.

What is the longest string of consecutive numbers you can make, using four numbers exactly once each time and using any of the four operations?

 Ask each pair to choose four different one-digit numbers. What is the longest string of consecutive numbers they can now make? To maintain the momentum of the challenge, stop the pairs after 10 minutes and ask them to share with the rest of the class the strategies, if any, they have found for generating the answers.

 Allow children to continue with the challenge.

Drawing out using and applying

 Invite pairs of children to say their four one-digit numbers and to explain how many consecutive numbers they were able to make in a string. Share one or two of the calculations they made to make some of their numbers.

What four one-digit numbers did you choose?

How many consecutive numbers were you able to make?

Tell me how you made the number ...?

Did any other pair choose the same four one-digit numbers?

How many consecutive numbers were you able to make?

Which numbers were easy to make? Why? Which numbers were more difficult? Why do you think this was?

What was a good number to include as one of your four one-digit numbers? Why was this a good number?

What was the longest unbroken string of consecutive numbers anyone found?

What strategies did you use for making your string of numbers as long as possible?

Assessing using and applying

- Children can randomly generate numbers and then organise them into consecutive strings.
- Children can initially work randomly, find a string of consecutive numbers and search for others to extend the string.
- Children can develop logical systems for generating consecutive or near numbers: for example, realising that adding, subtracting or multiplying by 1 (for example, $8 - 7$) can produce three consecutive numbers.

Supporting the challenge

- Suggest that two of the one-digit numbers the children choose are 1 and 2.
- Make sure children have number lines and number grids to help them.

Extending the challenge

- Children only use one digit repeated four times and the four operations. For example:
$1 = (4 \times 4) \div (4 \times 4)$;
$2 = (4 \div 4) + (4 \div 4)$; and so on.
- Children use each digit exactly once and each operation no more than once in each calculation.

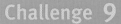

Domino puzzle

Using knowledge of addition and subtraction number facts to 20

Using and applying

Solving problems

Solve problems involving whole numbers

Reasoning

Explore patterns, properties and relationships and propose a general statement involving numbers; identify examples for which the statement is true or false

Maths content

Knowing and using number facts

- Use knowledge of place value and addition and subtraction
- Use knowledge of rounding, place value, number facts and inverse operations to estimate and check calculations

Key vocabulary

addition, subtraction

Resources

- Individual whiteboard and pen (for each pair)

For each child:

- RS14
- Squared paper

For *Supporting the challenge*:

- RS15
- Sets of dominoes

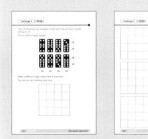

What other magic squares can you make, using eight dominoes?

Introducing the challenge

 Introduce/Reintroduce the idea of a magic square to the children. Display the following magic square:

8	16	9
12	11	10
13	6	14

Explain to/Remind the children that in a magic square, the sum of each column, row and diagonal is the same: this is the magic number.

What is the magic number for this magic square? (33)

 Write the following magic squares on the board and provide each pair of children with a whiteboard. Children complete each of the magic squares, using the numbers 1 to 9.

	7	
1		
	4	

	1	
		7
		2

 Once they have completed the magic squares, ask the children for the answers and the strategies they used to identify the missing numbers in the squares.

Can you tell me the missing numbers in this magic square?

How did you work out what they were?

How did knowing what the magic number is help you complete the magic square?

The challenge

 Provide each child with a copy of RS14 and either squared paper or RS15. Discuss the magic square made from dominoes with the children. Working with dominoes to solve a domino magic square is difficult because children have to introduce two numbers simultaneously while taking account of the rows and the columns.

Point out to the children that unlike the magic squares they were just working on, Domino magic squares do not have their diagonals totalling to the same amount.

This challenge needs much trial and improvement, and some systematic working and logical thinking. Encourage the children by asking questions such as:

What have you tried already?

Why didn't that work?

What if you kept that one there and tried other ones next to it?

Children can keep track of what they have tried before either using RS15 or by inventing their own system of notation on squared paper.

Drawing out using and applying

 When the children have found solutions, bring them back together for a whole-class discussion and ask them to share their strategies.

Who made another domino magic square, using eight dominoes? Show us your domino magic square.

Explain to us how you went about making it.

Did anyone make a different domino magic square, using the eight dominoes?

How did you go about making your domino magic square?

What is the same or different about these domino magic squares?

What was a good method of keeping track of what you tried?

Who used a different method? Can you explain it to us?

If you had to do a puzzle similar to this again, what would you do differently? Why?

Assessing using and applying

- Children can use trial-and-improvement methods effectively, such as: "I've tried that in each position and it doesn't work, so I'll put it with those others that don't work."
- Children can adjust what they have done as they go along. For example: "I need a 2-1 domino here, but I've used it there, so I'll see if I can use two other dominoes with 2 and 1 on them instead."
- Children can develop general logical strategies such as: "I'll try low numbers in the corners and higher numbers round the edges."

Supporting the challenge

- Start off by challenging the children to use eight dominoes and make the total in each row the same.
- Suggest the position of two or three dominoes to get the children started.

Extending the challenge

- The children can turn their invented domino magic squares into puzzles for each other by showing the position of any two dominoes somewhere in the puzzle and asking friends to find the correct place for the rest.
- Investigate different kinds of magic squares – there are many possibilities. For example, children complete a 4 by 4 square where each row, column and diagonal add to 10, using just the numbers 1 to 4.

Almost magic

Adding more than two numbers mentally

Using and applying

Reasoning

Explore patterns, properties and relationships, and propose a general statement involving numbers; identify examples for which the statement is true or false

Communicating

Explain reasoning, using diagrams and text; refine ways of recording, using images and symbols

Maths content

Knowing and using number facts

- Use knowledge of place value and addition and subtraction
- Use knowledge of rounding, place value, number facts and inverse operations to estimate and check calculations

Key vocabulary

addition, subtraction

Resources

For each child:
- RS16, pencil and paper

For *Supporting the challenge*:
- Calculator

For *Extending the challenge*:
- RS17
- Scissors

Can you sort out which numbers have been moved?

Introducing the challenge

 Children will need to do Challenge 9: Domino puzzle before doing this challenge. Reintroduce the idea of a magic square to the children. Display the following magic square:

8	16	9
12	11	10
13	6	14

Remind the children that in a magic square, the sum of each column, row and diagonal is the same: this is the magic number.

What is the magic number for this magic square?

The challenge

 Give each child a copy of RS16 and pencil and paper and introduce the challenge to the children. Explain that this is a 6 by 6 magic square, but that some of the numbers have been swapped around.

Can you sort out which numbers are in the wrong place and position them correctly?

 Encourage the children to work as a group, apportioning parts of the challenge and pooling their results. You may want to work with the groups to help them decide how they are going to split up the challenge. Suggest that at least two children calculate each row and column to check for errors.

Drawing out using and applying

 As well as discussing with the children how they found the correct position of all the numbers, discuss how they went about the challenge.

What numbers are in the wrong positions? Where should they be?

How did you start out to solve the problem?

How did it help to know what the magic number was?

How did you find the magic number?

As a group, how did you organise yourselves?

If you were to do the challenge again, how would you organise yourselves differently?

Assessing using and applying

- Children can work on the challenge without much regard for how the others in the group are contributing.
- Children can work on their own initially, but then share their results to check that the work is being covered.
- Children can talk through together how best to solve the problem and plan accordingly.

Supporting the challenge

- Children use a calculator.
- Tell the children that the magic number for the rows and columns is 111, to start them off.

Extending the challenge

- Pairs move around other numbers in the correct square as a challenge to others.
- Using RS17, pairs of children play 'Fifteens'. Children take turns to place one of the digits from 1 to 9 onto the 3 by 3 grid. The winner is the first player to create a row, column or diagonal that totals 15.

Factors

Identifying pairs of factors of numbers to 100

Using and applying

Reasoning

Explore patterns, properties and relationships, and propose a general statement involving numbers; identify examples for which the statement is true or false

Communicating

Explain reasoning using diagrams and text

Maths content

Knowing and using number facts

- Identify pairs of factors of two-digit whole numbers
- Recall quickly multiplication facts to 10×10; derive quickly corresponding division facts

Key vocabulary

factor, multiple, common multiple, multiplication, division

Resources

- Individual whiteboard and pen (for each pair)
- RS18 (optional, for each child)

For *Supporting the challenge*:

- Multiplication square (see RS19)

Which two-digit numbers have the most factors?

Introducing the challenge

 Write a two-digit number on the board (for example, 15) and remind the children of the meaning of the word 'factor'. (A factor is a whole number which will divide exactly into another whole number.)

Children say all the factors of that number. As they offer the numbers, make a list on the board of the pairs of factors in a systematic way, starting with 1:

$$1 \times 15$$

$$3 \times 5$$

Who can tell me a pair of factors for 36?

Who can tell me another pair?

Are there any more?

Explain to the children how working in a systematic way and making a list helps to make sure that they have found all of the factors.

Ask the children to explain why some numbers are not factors of 15: for example, 6 (because 6 does not divide evenly into 15).

 Write another two-digit number on the board: for example, 36. Children work together to write down all the factors for that number on their individual whiteboard.

If appropriate, repeat the above for another two-digit number.

Identifying patterns and relationships
and explaining reasoning

The challenge

 Introduce the challenge to the children, using RS18. Remind the children to use their knowledge of the multiplication facts to 10×10 and the related division facts to help them in identifying factors. Some children may find it helpful to use a multiplication square for this (see RS19).

As the children work through the investigation, ask them to think carefully about the best way to keep a record of each of the numbers they try and their factors. Tell them to look out for any patterns and relationships they notice.

What kind of two-digit numbers would be good to start with? Why?

How many factors does this number have?

Can you think of another number that might have more factors than this? Why do you think it might have more factors?

How are you keeping track of the numbers to find factors for?

Do you see a pattern emerging for these numbers? What do they all have in common?

Drawing out using and applying

 Ask the children to identify the two-digit numbers with the most factors. Discuss their methods of working and any patterns and relationships they noticed.

Which two-digit numbers have the most factors?

How did you work it out?

Who chose numbers randomly? Who didn't? Why did you choose the numbers you did?

How did you keep a record of the number you chose and their factors? Did this help you to identify other numbers to try? How?

What can you say about all the two-digit numbers with 12 factors?

Assessing using and applying

- Children can choose two-digit numbers and find their factors, but are haphazard in their choice.
- Children can use their knowledge of multiplication and division facts to choose which numbers to try and find their factors and then use these to find other similar numbers.
- Children can see a relationship between some of the numbers with 12 factors and use this to predict other numbers with 12 factors.

Supporting the challenge

- Provide the children with a multiplication square (see RS19). This will help them find factors for some two-digit numbers, and from this they may begin to recognise patterns and relationships and identify other two-digit numbers with a large number of factors.
- Suggest that children find factors for numbers greater than 50.

Extending the challenge

- Which three-digit numbers have the most factors?
- Which two-digit numbers have the fewest factors?

Common multiples

Recognising common multiples

Using and applying

Reasoning

Explore patterns, properties and relationships and propose a general statement involving numbers; identify examples for which the statement is true or false

Communicating

Explain reasoning using diagrams

Maths content

Knowing and using number facts

- Find common multiples
- Recall quickly multiplication facts to 10 × 10; derive quickly corresponding division facts

Key vocabulary

multiplication, multiple, common multiple, relationship, generalisation

Resources

- NNS ITP: Number grid

For each child:

- RS20
- Coloured pencils

For *Supporting the challenge*:

- Multiplication square (see RS19)

> # Which numbers between 1 and 100 are multiples of 4 as well as multiples of 6?

Introducing the challenge

 Display the NNS ITP: Number grid. Adjust the base multiple to 3. Remind the children of the meaning of the word 'multiple'. (A multiple is a number that can be divided into another number.) Children suggest the first 10 multiples of 3. As they do this, highlight these numbers on the 100-grid.

Next, children predict several more multiples of 3 and identify the pattern.

What do you notice about all these multiples of 3?

How can you use this to predict other multiples of 3?

Tell me a number with 5 tens that is a multiple of 3?

Reveal all the multiples of 3 from 3 to 99.

What patterns do you notice?

What relationships can you see between these numbers?

If appropriate, reset the ITP and choose another number. Ask the children to suggest multiples, comment on patterns and relationships and make predictions about other multiples to 100.

1	2	3	4	5	6	7	8	9	10
11	12	13	14	15	16	17	18	19	20
21	22	23	24	25	26	27	28	29	30
31	32	33	34	35	36	37	38	39	40
41	42	43	44	45	46	47	48	49	50
51	52	53	54	55	56	57	58	59	60

The challenge

 Give each child a copy of RS20 and introduce the challenge to the class. Discuss the meaning of the term 'common multiple'. Remind the children to use their knowledge of the multiplication facts to 10 × 10 to help them identifying multiples. Some children may find it helpful to use a multiplication square for this.

As the children work through the investigation, ask them to look carefully at the patterns that emerge on the 100-grid and use this to predict other multiples and common multiples. Children look out for any patterns and relationships they notice and make a generalisation about the multiples common to both numbers.

Tell me some of the multiples of 4. How do you know this?

What other numbers are common multiples of 4 and 6?

Can you use this to predict other common multiples of 4 and 6?

Without using the 100-grid, how are you keeping a record of the common multiples?

Do you see a pattern emerging for these numbers? What do they all have in common?

Drawing out using and applying

 Children identify some of the multiples of 4 and 6 and their common multiples. As they say these, highlight the numbers on the ITP. Eventually, ask the children to predict two-digit multiples of 4 and 6 and their common multiples.

Discuss with the children the pattern that occurs for all the common multiples of 4 and 6.

1	2	3	4	5	6	7	8	9	10
11	12	13	14	15	16	17	18	19	20
21	22	23	24	25	26	27	28	29	30
31	32	33	34	35	36	37	38	39	40
41	42	43	44	45	46	47	48	49	50
51	52	53	54	55	56	57	58	59	60

What are the next common multiple of 4 and 6 after 96? How do you know this?

What can you tell me about all the multiples of 4 and 6?

Finally, children say two numbers from 2 to 10 and their common multiples.

What do you notice about each of these numbers? What do they have in common?

How did you work out the common factors without using the 100-grid?

Assessing using and applying
- Children can identify common multiples of two numbers from 2 to 10, to the tenth multiple.
- Children can identify common multiples of numbers 2 to 10, to the tenth multiple, and use these to make predictions of other two-digit multiples.
- Children can make a generalisation of common multiples.

Supporting the challenge
- Provide the children with a multiplication square (see RS19) to help them work out the multiples of numbers 2 to 10, to the tenth multiple.
- Encourage the children to use the 100-grid to help them identify patterns between common multiples.

Extending the challenge
- What about common multiples of 2, 3 and 5?
- Investigate the meaning of the term 'Lowest Common Multiple' (LCM). When is it useful to know this? Provide some examples.

In a year ✓

Making estimations and calculating, using the four operations

Using and applying

Enquiring

Plan and pursue an enquiry; present evidence by collecting, organising and interpreting information; suggest extensions to the enquiry

Communicating

Explain reasoning using diagrams, graphs and text; refine ways of recording using images and symbols

Maths content

Knowing and using number facts

- Use knowledge of rounding, place value, number facts and inverse operations to estimate

Calculating

- Use a calculator to solve problems, including those involving decimals; interpret the display correctly

Key vocabulary

estimation, approximation, calculation

Resources

For each group:
- RS21
- Calculator

For each child:
- Pencil and paper

What do you do in a year?

Introducing the challenge

 Arrange the children into groups. Provide each group with a copy of RS21 and a calculator.

 Discuss with the children the different quantities listed on the resource sheet. Tell the groups that they have five minutes to add as many other questions to the list as they can.

What other 'in a year' questions can you think of?

 After the five minutes, ask each group in turn to suggest one of the items on their list. Write up the suggestions on the board.

 Give each group two minutes to decide on one of the items to estimate 'a year's worth of'. Record each group's choice on the board. Point out that it is okay to use one of the questions that was already on the resource sheet or to select something that somebody else suggested.

As a group, I want you to decide on one 'in a year' question.

Why did you decide upon that question?

What are you going to have to do to find an approximate answer to your question?

The challenge

 Set the children off to estimate the total by generating some data to work on. Encourage them not to simply make guesses at the answers but to use what they know to make sensible estimates. For example, they may know how long it takes to walk to school, but not know the distance, so they may decide to walk around the playground for that length of time to calculate the distance from home.

What do you know already that can help you answer your question?

What do you need to do to answer this question?

How are you going to find that out?

Drawing out using and applying

 Ask each group to talk through the method they used to arrive at their estimate. Emphasise that the other children should listen carefully and make suggestions for ways in which the estimate might be improved.

What was your question?

What is your approximate answer?

How did you arrive at your estimate?

If you had to investigate this question again, what would you do differently? Why?

Assessing using and applying

- Children take little account of their audience when explaining.
- Children can cover all the main points of explanation but assume that the audience understands without checking if they are indeed following.
- Children can make their method clear to children who are not working on the same task.

Supporting the challenge

- Work with the children breaking down the problem into simple steps: for example, looking at how far they walk to school in a week.
- Encourage the children to use a calculator when calculating with large numbers.

Extending the challenge

- Ask the children to make other 'In a year' questions that are not related to their own experiences but perhaps to their parents instead. Groups investigate one of these questions.
- Children try working backward: How long would it take to walk 1000 miles?

Four-digit number

Calculating and exploring properties of numbers

Using and applying

Representing

Represent a puzzle by identifying and recording the information or calculations needed to solve it; find possible solutions and confirm them in the context of the problem

Reasoning

Explore patterns, properties and relationships and propose a general statement involving numbers; identify examples for which the statement is true or false

Maths content

Calculating

- Extend mental methods for whole-number calculations
- Use efficient written methods to add, subtract, multiply and divide whole numbers

Key vocabulary

number, digit, add, subtract, multiply, divide

Resources

For each pair:

- Individual whiteboard and pen (optional)
- RS22 (optional)

For *Supporting the challenge*:

- Number lines and grids

What questions can you ask about your number?

Introducing the challenge

 Introduce the challenge to the class by asking for a four-digit number and writing it on the board: for example, 1542. Ask children to think of questions using the digits in the number. Explain that the questions need to be related to the number or the digits in the number and involve either properties of numbers or calculations.

Use the digits to make numbers between 100 and 500.

What do you need to add to 1542 to make a multiple of 50?
Use the digits to make numbers that are multiples of 5.

What is the number nearest to 1542 that is a multiple of 9?
Use the digits to make a calculation where the answer is 90.

Use all four digits to make a division calculation with an answer of 38.

Encourage the children to think flexibly.

 Ask the children to spend a few minutes to think of different questions to ask, working in pairs. Explain that they need to know the answer to their question. If appropriate, children record their question and answer on individual whiteboards.

 After enough time, bring the class back together and ask pairs of children to ask their question to the rest of the class.

The challenge

 If appropriate, give each pair a copy of RS22. The children work in pairs to invent the numbers and the questions. They swap with another pair and answer those questions. They then swap back to check each other's solutions. On occasion, it is likely that those children asking the questions will not have thought of all the solutions. As children work, ask them questions that encourage them to think of all possible solutions.

What other properties of those numbers could you think of?

What if you turned that number round the other way?

Does looking at the units digit give you any idea?

What operations have you not used yet?

Drawing out using and applying

 Ask a pair of children to ask one of their questions. Another pair of children gives the answer and explains their working. Repeat the exercise if time permits.

Tell us one of your questions. Who had to answer that question?

What is the answer to this question? How did you work it out?

Is the answer correct? Did you work out the answer in the same way? How did you work out the answer to your problem?

Finally, share ideas with the class about good questions to ask and challenging questions to work on.

Assessing using and applying

- Children can invent interesting questions that show a knowledge of the relationship between numbers.
- Children can discuss properties of numbers with each other, building on each other's ideas and challenging each other's thinking.
- Children can present their ideas to each other and evaluate them fairly.

Supporting the challenge

- Children think of a three-digit number.
- Make sure children have number lines and other equipment available when they are devising and answering questions.

Extending the challenge

- Children think of a five-digit or six-digit number.
- Children include a decimal point: for example, 000.0, 00.00 or 0.000.

Float for the stall

Adding money and using coins to make a given total

Using and applying

Solving problems

Solve problems involving whole numbers and decimals

Communicating

Explain reasoning using diagrams and text; refine ways of recording using images and symbols

Maths content

Calculating

- Extend mental methods for whole-number calculations
- Use efficient written methods to add and subtract whole numbers and decimals with up to two places

Key vocabulary

addition, subtraction, float, money, coin, amount, prices, cheapest, most expensive

Resources

- RS23 (optional, for each pair)

For *Supporting the challenge*:

- Selection of coins in all denominations

What coins will you ask for?

Introducing the challenge

 Discuss the idea of a 'float' with the children and find out their experiences of this. Ask them to suggest reasons why a float is necessary. They may have seen till operators in supermarkets emptying out money into the till.

What amounts might be in the bag?

The challenge

 Introduce the challenge to the children, using RS23. Ask for ideas about what to take into consideration in tackling this problem, such as the price of any item (and hence the range of possible change needed for the first purchase), and the price and number of cheaper items.

 As the children work on the challenge, look out for points for discussion and, where appropriate, draw the children together to look at useful strategies for tackling the problem.

What if I buy something for £2.63 and offer you £10?

What change will you give me?

What coins would you use?

Drawing out using and applying

 Compare the different ways the children have solved the problem and ask the children to justify their solutions. Are all the solutions equally workable? Encourage the children to evaluate the strategies used.

Who would like to describe to the rest of the class how they solved the problem?

Who did it a different way?

Tell us what coins are in your float.

Who has got a different combination of coins?

Are there any other combinations of coins that people had?

How come we have so many different possibilities? Can they all be right?

Which combinations of coins are better than others? Why?

Assessing using and applying

- Children can decide on the important factors in the challenge, such as analysing the coins most needed and the quantity that would be most useful.
- Children can calculate the amounts needed in a systematic way.
- Children can show clearly what decisions they have made and what conclusions they have reached.

Supporting the challenge

- Children use coins to help them work out possible combinations. Start with a smaller float and simpler amounts such as £10 and 10p and £1.25.
- Help children record their work systematically to keep track.

Extending the challenge

- Children invent an amount that the stall took at the fair and work out how to bag it up sensibly to take it to the bank.
- Using their float, children work out the maximum number of items that the first customers could buy with notes instead of coins, so that they are left with no change at all.

How much do you eat?

**Making estimations and approximations;
solving calculations involving money**

Using and applying

Solving problems

Solve problems involving
whole numbers and decimals
and all four operations, choosing
and using appropriate calculation
strategies, including calculator
use

Enquiring

Plan and pursue an enquiry;
present evidence by collecting,
organising and interpreting
information; suggest extensions
to the enquiry

Maths content

Knowing and using number facts

- Use knowledge of rounding,
  place value, number facts and
  inverse operations to estimate
  and check calculations

Calculating

- Use efficient written methods
  to add, subtract and multiply
  whole numbers and decimals
  with up to two places

Key vocabulary

money, estimation,
approximation, calculation,
addition, multiplication

Resources

- RS24 (optional, for each child)

For *Supporting the challenge*:

- Calculators,
  shopping
  receipts,
  catalogues

> **How much does it cost to feed you for a day?
> For a week? For a month? For a year?**

Introducing the challenge

 Discuss what the children typically eat in one day. List some of
the items on the board and encourage the children to suggest
some realistic prices for them. Discuss with them how you
cannot buy some things in a single-portion size: for example,
a glass of milk.

Tell me some of the food that you eat in a day.

Approximately how much does this cost?

Do you drink the entire carton of milk yourself?

Can you estimate how much of the milk you use?

How will you estimate how much a portion would cost?

Make sure the children account for the different times
throughout the day they eat something, including any snacks
they may have.

The challenge

 Introduce the challenge to the children, using RS24. Discuss
what timescale they are all going to work on. For example,
for some foods, it might make more sense to estimate how
much children eat in a week or month than in a day. Encourage
the children to share their ideas and to check that they are
thinking of reasonable prices and answers.

Do you eat a whole packet of cereal every day?

Approximately how long does a box of cereal last?

How many people in your family have cereal every morning?

Can you estimate how much you eat in a week? In a month?

Once the children have spent time working individually identifying what food they eat and allocating approximate prices to each of these, they work in pairs to briefly discuss and compare the lists. This will help all children make sure that they have not forgotten particular food items they eat in a day. It will also provide an opportunity for them to make comparisons between the prices they have allocated to different food items and ensure that their prices are realistic.

Drawing out using and applying

 Invite individual children to read through some of the food items on their list and the prices they allocated to each.

Whose list is similar to Gavin's?

Are your prices similar? How do they differ?

> Individual children say how much they estimate it costs to buy all the food they eat.

How much does it cost to feed you for a day?

What about for a week? For a month? For a year?

How did you work out how much it costs to buy the food you eat in a day?

Did you always calculate the cost per day?

Did anyone calculate the cost of some items per day, others per week or per month? Explain to us how you did this? Why did you do this?

What was the largest amount of money that anyone calculated? What could this mean?

What was the smallest amount of money that anyone calculated? What might this mean?

How big is the range between these two?

> Finally, list all the children's estimates on the board and ask the children to calculate the mean average (total of all the values divided by the number of values) and median (middle value when all the values have been ordered smallest to largest).

Which of these do you think gives the better estimate? Why?

Assessing using and applying

- Children need a lot of support in deciding on appropriate prices.
- Children can work with one unit of time: for example, everything they eat in a day.
- Children use flexible approaches to the challenge and discuss their various strategies.

Supporting the challenge

- Provide the children with shopping receipts and catalogues to help them decide the cost of items of food.
- Children work out the costs for 1 day, 10 days, 30 days.

Extending the challenge

- How much does it cost to keep a pet?
- What else do the children need money for?

Heartbeat

Using a calculator to calculate with large numbers

Using and applying

Enquiring

Plan and pursue an enquiry; present evidence by collecting, organising and interpreting information; suggest extensions to the enquiry

Communicating

Explain reasoning using diagrams and text; refine ways of recording, using images and symbols

Maths content

Knowing and using number facts

- Use knowledge of rounding, place value, number facts and inverse operations to estimate and check calculations

Calculating

- Use a calculator to solve problems; interpret the display correctly

Key vocabulary

calculate, calculator, time, second, minute, hour, day, week, month, year, lifetime

Resources

For each pair:
- RS25 (optional)
- Stopwatch or watch with second hand
- Calculator

How many times does your heart beat in a day?

Introducing the challenge

 Talk with children about pulse rate and how it relates to the beating of the heart. Discuss how the heart beats at different rates depending on what you are doing. Each time the heart beats, it pushes blood through the arteries which you can feel as a pulse. To find out how fast your heart is beating, you need to take your pulse rate. The best way to do this is to place two fingers (not a thumb, as this has a pulse) on the artery near the surface of the skin on the inside of the wrist.

 Show the children how to take each other's pulse over a 30-second period.

Who has the quickest pulse? The slowest?

What does this mean?

The challenge

 Provide each pair of children with a stopwatch or watch with second hand, a calculator and a copy of RS25. The activity is self-explanatory.

Encourage the children to set out their working clearly so that someone else would be able to follow it through. Can they find a different way of doing the calculation in order to check their results?

Think about how you are going to record your results so that someone else can read and understand them.

What information is important to know when working out how many times your heart beats in a day?

Drawing out using and applying

 Ask the children to swap their working out with someone in the class. This child should not be the same child they have worked with on the first part of the challenge. Give children time to read through and understand the presentation that they have been given.

 Invite individual children to explain the method used in the presentation they were given. Can the child explain clearly how the other person worked it out?

Harry, will you explain to the rest of the class how Justine went about finding how many times her heart beats in a day. What about in a week? A month?

Once different children have explained other children's results, discuss the results.

Who has the quickest pulse? Who has the slowest?

What does this mean?

How much variation is there between the different answers?

Why might this be so?

What is the average pulse rate for this class?

How accurate do we need to be in finding out about pulse rates?

Assessing using and applying

- Children can present their results only for themselves; their method is not clear to others.
- Children can explain their method clearly.
- Children can explain a way of checking their results, using a different method.

Supporting the challenge

- Children measure how many times their heart beats in one minute.
- Provide the children with a list showing the relationship between units of time. For example;

$$30 \text{ seconds} = \tfrac{1}{2} \text{ minute}$$
$$60 \text{ seconds} = 1 \text{ minute}$$
$$60 \text{ minutes} = 1 \text{ hour}$$
$$24 \text{ hours} = 1 \text{ day}$$
$$7 \text{ days} = 1 \text{ week}$$

Extending the challenge

- How many times does your heart beat in a year? In 10 years? In 30 years?
- Children see how they can vary their pulse rate by jumping up and down vigorously, and lying still, and work out the range of pulses they can have in one minute.

Sports gear

Calculating with decimals in the context of money

Using and applying

Solving problems

Solve problems involving whole numbers and decimals and all four operations, choosing and using appropriate calculation strategies, including calculator use

Enquiring

Plan and pursue an enquiry; present evidence by collecting, organising and interpreting information; suggest extensions to the enquiry

Maths content

Calculating

- Use efficient written methods to add, subtract and multiply whole numbers and decimals with up to two places
- Use a calculator to solve problems, including those involving decimals; interpret the display correctly

Key vocabulary

money, pounds, pence, total, cost

Resources

For each child:

- Individual whiteboard and pen
- RS26
- RS27
- Pencil and paper, calculator

What is the total cost of the sports gear for your school team?

Introducing the challenge

 Explain that the challenge involves kitting out a sports team with sports gear with names, numbers and badges.

 Write up prices of T-shirts, letters and numbers: £5.95 for a T-shirt, 35p for each letter, 45p for each digit. Children work out the cost of putting their name and favourite sport number on a T-shirt, using their whiteboard.

 Invite a pair to be a 'team', put their costs together and write the amount on the board.

Invite the class to ask questions to find out how they have arrived at those costs. Write up the findings.

Gather ideas about the possible teams for different sports and the numbers of children involved.

Football	Rugby	Basketball	Netball	Rounders/ Softball	Cricket
11 players	15 players	5 players	7 players	9–12 players	11 players

You could change the context to non-sport school teams such as a singing group, a brass band, a dance troupe or a chess team, depending on the children's interests.

The challenge

 The challenge requires the children to make several decisions about prices, names and numbers, colours and sizes. They need to fill in this information in the correct place on RS26 and then use RS27 for each team member. Children need to do a separate calculation for the names and numbers. The challenge involves sorting all this information, and children need to use paper for working out individual calculations.

When they have completed the challenge, children work in pairs, calculating each other's total cost of the sports gear for their school team.

When they have done this, they can become sportswear suppliers' order form checkers. They work in pairs and check the two order forms and total cost of another pair.

Drawing out using and applying

 Ask children to talk about their school team, the number of children involved, the cost per child and the total cost for the team.

What sport does your team play?

How many players are there in the team?

Did you include additional kits for substitute players?

What is the cost per team member?

What is the total cost of the sports gear for your school team?

Encourage comparisons between the same sports.

Who else bought kits for a football team?

How much did each kit cost?

What is the total cost of your team's sports gear?

Is there a big difference between the cost for your team and the cost for Leo's team? Is this difference acceptable?

At the end of the challenge, discuss with the class the kind of errors that children are likely to make on the order form. Suggest that it would be a good idea to double-check when filling in the form.

Assessing using and applying

- Children can work out all the information they need from the different parts of the catalogue.
- Children can collect together this information and fill in the form systematically and accurately.
- Children can develop systems for double-checking their information entered on the order form and the accuracy of the calculations they have made.

Supporting the challenge

- Simplify the problem to just T-shirts and hats, to reduce the number of variables.
- Children work in pairs, choosing the same sport and working together to complete the order forms.

Extending the challenge

- Set a budget for a team, with more choices about what items team members should have. Children can group in a 'team' and work together on the choices involved and decisions to be made.
- Children invent their own class sports catalogue, including sports equipment as well as sportswear.

Three in a line

Estimating and calculating, including decimals and negative numbers

Using and applying

Representing

Represent a problem by identifying and recording the information or calculations needed to solve it; find possible solutions

Reasoning

Explore patterns, properties and relationships; identify examples

Maths content

Calculating

- Extend mental methods for whole number and decimals with up to two places
- Use a calculator to solve problems, including those involving decimals; interpret the display correctly

Key vocabulary

add, subtract, multiply, divide, whole number, decimal, negative number

Resources

For each pair:
- RS28
- Calculator
- Counters in two different colours
- Two 1–6 dice

For *Extending the challenge*:
- 0–9 or 1–12 dice

How will you use the two numbers on the dice to make one of the numbers on the grid?

Introducing the challenge

 Introduce the game to the children, you playing against the class. Roll two 1–6 dice and ask the children to suggest the different calculations they could make using those two numbers.

$$3 + 1 = 4 \quad 3 - 1 = 2 \quad 1 - 3 = -2 \quad 3 \times 1 = 3 \quad 1 \div 3 = 0.33 \quad 3 \div 1 = 3$$

Ask the children to explain their methods.

Focus on a decimal number such as 0.5 and ask children to offer all the pairs of numbers on the dice that will give 0.5: $1 \div 2, 2 \div 4, 3 \div 6$

The challenge

 Children play the game on RS28 in pairs.

As the children play, ask questions such as:

How will you use the two numbers on the dice to make one of the numbers on the grid?

You have rolled 6 and 4. What are all the different numbers you can make?

What calculations have you tried so far? Are there any you haven't tried?

If you want that number next, what calculations would help?

What numbers do you need to roll?

Are there any numbers on the grid you can get only one way?

Which pair of numbers gives you the most possible answers?

Which pair of numbers gives you the least possible answers?

Drawing out using and applying

 Discuss all the above questions with the class, asking children if they have come to any general conclusions such as how to get 0.5, 0.25 and 0.75, or how to get negative numbers. Expect the children to be as coherent in their reasoning as possible.

Assessing using and applying

- Children can use their knowledge of numbers flexibly to try out different ideas and try all possible calculations.

- Children can say what calculations they are likely to need, such as: "That's got to be a subtraction"; "You have to divide to get that."

- Children can discuss general ideas such as: "To get a whole number and a decimal, you have to divide a larger number by a smaller"; "To get a negative number, you have to subtract a larger number from a smaller number."

Supporting the challenge

- Do not use RS28. Instead, each child writes down the numbers 0 to 12 on a sheet of paper. They play the game as described on the resource sheet, aiming to be the first player to cross off all the numbers from 0 to 12.

- Play the game as described above, including negative numbers in the range: for example, the numbers from − 5 to 10.

Extending the challenge

- Children use more dice or dice with higher numbers, such as a 0–9 or 1–12 dice.

- Children invent their own board and test and modify it to make sure it works.

Magic 999

Using written methods to add and subtract whole numbers

Using and applying

Representing

Represent a problem by identifying and recording the information or calculations needed to solve it; find possible solutions and confirm them in the context of the problem

Reasoning

Explore patterns, properties and relationships, and propose a general statement involving numbers; identify examples for which the statement is true or false

Maths content

Knowing and using number facts

• Use knowledge of rounding, place value, number facts and inverse operations to estimate and check calculations

Calculating

• Use efficient written methods to add and subtract whole numbers

Key vocabulary

addition, subtraction, estimation

Resources

For each child:
• RS29
• Pencil and paper

For *Supporting the challenge*:
• Calculator

Can you subtract from 999 and then add to work out any subtraction calculation?

Introducing the challenge

 Write up 999. Invite children to suggest three-digit numbers to subtract from 999 and challenge the class to do the calculations as quickly as they can.

Why is this easy to do?

Talk about Pat, who will only subtract from 999. Show her method for calculating $653 - 487$.

She won't do $653 - 487$, but she will do $999 - 487$. Why do you think Pat will do $999 - 487$?

$$\begin{array}{r} 999 \\ -\ 487 \\ \hline 512 \end{array}$$

She likes adding, so she adds 653 to the answer.

$$\begin{array}{r} 512 \\ +\ 653 \\ \hline 1165 \end{array}$$

She crosses out the '1' at the far left of the answer (actually one thousand) and adds the '1' to the answer.

$$\begin{array}{r} 512 \\ +\ 653 \\ \hline 165 \\ +\quad 1 \\ \hline 166 \end{array}$$

Invite children to check whether the answer is correct. They can do the calculation any way they like to convince themselves.

The challenge

 Provide each child with a copy of RS29 and pencil and paper. Children create some more subtraction calculations involving pairs of three-digit numbers to check if this method always works.

Drawing out using and applying

 Children discuss the results of the subtraction calculations
they created.

Did anyone find a calculation that did not work?

Can anyone explain why this method works?

Can it be extended to four-digit numbers? How?

Do you think that this is a written method you might use?

When might you use it?

*Is it more efficient than the written method for subtraction
you currently use?*

Assessing using and applying

- Children can try out numbers randomly and are convinced it works on the basis of a few examples.
- Children can search for examples that they think will 'test' the method: for example, numbers with zeros in them.
- Children can use their knowledge of place value to try to understand why the method works.

Supporting the challenge

- Simplify the problem by using pairs of two-digit numbers such as 73 − 46 and subtracting from 99.
- Make sure children have a calculator to check their answers.

Extending the challenge

- What would happen if you subtracted the smaller number from 1000?
- What if you chose a pair of four-digit numbers and subtracted the smaller number from 9999?

Fraction cards

Finding fractions of numbers

Using and applying

Solving problems

Solve problems involving whole numbers, choosing and using appropriate calculation strategies, including calculator use

Representing

Represent a puzzle or problem by identifying and recording the information or calculations needed to solve it; find possible solutions and confirm them in the context of the problem

Maths content

Calculating

- Find fractions using division
- Use a calculator to solve problems, including those involving fractions

Key vocabulary

number, digit, fraction, multiplication, division

Resources

For each pair:

- RS30
- Set of 0–9 digit cards
- Pencil and paper

For *Supporting the challenge*:

- Multiplication square (see RS19)

How many different fraction statements can you make?

Introducing the challenge

 Display the digit cards 0, 2–9 in a row at the top of the board and underneath this the following arrangement:

| 0 | | 2 | 3 | 4 | 5 | 6 | 7 | 8 | 9 |

$$\frac{1}{\square} \quad \square\;\square \quad \square$$

Move four of the digit cards to make a true statement.

$$\frac{1}{3} \times \boxed{2}\,\boxed{4} = \boxed{8}$$

Explain to the children how you worked out the answer to one third of 24 (24 ÷ 3) and how you have used four of the digit cards to make a true statement.

Repeat the above, using the zero in the tens place:

$$\frac{1}{2} \times \boxed{0}\,\boxed{6} = \boxed{3}$$

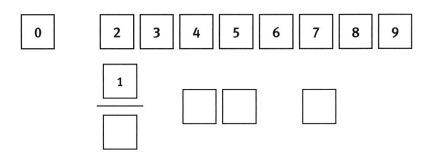

Ask one or two children to come to the board and arrange the cards to make other true statements.

How did you work out the answer to this calculation?

The challenge

 Arrange the children into pairs and provide each pair with a copy of RS30, a set of 0–9 digit cards and pencil and paper. Introduce the challenge to the children.

How many different fraction statements can you make?

Explain to the children that they do not have to use the digit cards to find the different fraction statements, but to use them if they feel they may be helpful.

How many statements can you make involving a half? What about a third?

Have you found all the statements involving a quarter? How do you know?

Drawing out using and applying

 Children suggest all the different unitary fractions that are possible. There are eight altogether ($\frac{1}{2}$, $\frac{1}{3}$, $\frac{1}{4}$, $\frac{1}{5}$, $\frac{1}{6}$, $\frac{1}{7}$, $\frac{1}{8}$ and $\frac{1}{9}$). Work systematically with the class for each fraction, starting with a half. Children discuss the different methods they used to ensure that they found all the different statements possible.

How many different fractions are possible?

How many statements could you make involving a half? Why couldn't you make any more?

Which fractions were you able to make the most statements for?

Who used the digit cards to help? How did they help you?

Who started to use the digit cards and then stopped? Why did you stop?

Who did not use the digit cards at all? What did you do?

Assessing using and applying

- Children can arrange the digit cards to make fraction statements and calculate the answers.
- Children can find all the different unitary fraction statements possible, using the digit cards to help them.
- Children can work systematically to find all the different fraction statements possible.

Supporting the challenge

- Remind children of the link between fractions of amounts and multiplication and division.
- Provide children with a multiplication square (see RS19).

Extending the challenge

- What if you chose five digit cards to complete this fraction statement?
- What about non-unitary fractions?

Find the missing alien

Visualising and describing shapes

Using and applying

Representing

Represent a puzzle by identifying and recording the information needed to solve it; find possible solutions and confirm them in the context of the problem

Reasoning

Explore patterns, properties and relationships and propose a general statement involving shapes; identify examples for which the statement is true or false

Maths content

Understanding shape

- Identify, visualise and describe properties of rectangles, triangles and regular polygons
- Use knowledge of properties to draw 2D shapes

Key vocabulary

two-dimensional (2D) shape, circle, triangle, square, rectangle, diamond, pentagon, hexagon, octagon, pattern, relationship, properties

Resources

- RS31

For each child:

- Individual whiteboard, pen, RS32

For *Extending the challenge*:

- Logic puzzle books

How do you know what the missing alien looks like?

Introducing the challenge

 Introduce the challenge to the children by displaying a different but similar puzzle to that on RS32. You may wish to use the example on RS31 and either display it on an interactive whiteboard or print out an enlarged copy of the puzzle.

 Explain to the children that one of the aliens is missing and you would like them to draw it on their individual whiteboards.

 Once the children think that they have successfully identified the missing alien, they show their picture of the missing alien to their partner and explain why they think this is what it looks like.

 After enough time, bring the whole class back together and ask the children for the solution to the puzzle as well as a full justification for their solution.

What shape are the eyes of the missing alien?

Describe the alien that is missing. How do you know?

Could it be anything else? Why not?

Ask a child to draw the missing alien in the blank square:

The challenge

 Provide each child with a copy of RS32 and talk through it with the class. Children identify the missing alien on RS32 and then make up some similar shape logic puzzles for a friend to solve.

How do you know it is going to be that shape?

Why have you drawn that many legs?

Is there a pattern that will help you?

When the children invent their own puzzles, they also need to create a logical framework for each one. Ask the children to make this explicit to you as they work on it.

 The children can swap their puzzles with a partner and solve each other's.

Drawing out using and applying

 Individual children identify the missing alien on RS32 and explain how they worked it out.

Can you describe the missing alien?

How do you know it has these features?

Did anyone identify a different alien?

Children choose one of the puzzles they created and write a description of the logical connections within it.

Individual children present their description to the rest of the class. The class can challenge if they don't understand, and the partner who completed the puzzle can comment on their own understanding of the logic.

Who would like to describe one of their logic shape puzzles to the class?

How did you go about creating your puzzle?

How did your partner identify the missing alien?

Conclude by compiling a list of strategies that are useful for solving logic problems, such as counting the numbers of shapes in each row and column.

What did you do to help you find out the missing alien?

Did anyone use a different strategy?

What else did you do?

Assessing using and applying

- Children can see a pattern of connections in the puzzle.
- Children can explain the logic behind the sequence of shapes.
- Children can create a logical system for inventing new puzzles.

Supporting the challenge

- Make the puzzles easier by having fewer items.

⊙ ⊙	⅋ ⅋	● ●
⚁	◆	¥
U	∧	∧
● ●	⊙ ⊙	⅋ ⅋
◆	¥	⚁
∧	∧	U
⅋ ⅋	● ●	⊙ ⊙
¥	⚁	◆
∧	U	∧

- Help children work out the different attributes, then start them off by asking leading questions such as: "How many different-shaped aliens are there?": "Look at the aliens' legs. How are they different?"; "What do you notice about each of the aliens that have three legs?"

Extending the challenge

- Make the puzzles more complex by adding more items, by including transformations such as rotations and reflection, by doubling and halving items, and so on.
- Analyse visual logic puzzles of this kind found in puzzle books and work out strategies for solving them.

Front and sides

Interpreting 2D representations of 3D solids and visualising movements of shapes

Using and applying

Representing

Represent a problem by identifying and recording the information needed to solve it; find possible solutions and confirm them in the context of the problem

Reasoning

Explore patterns, properties and relationships, and propose a general statement involving shapes; identify examples for which the statement is true or false

Maths content

Understanding shape

- Identify, visualise and describe properties of rectangles, triangles, regular polygons and 3D solids; use knowledge of properties to draw 2D shapes
- Visualise and draw the position of a shape after a movement

Key vocabulary

two-dimensional (2D), three-dimensional (3D), cube, smallest, arrangement

Resources

For each child:
- RS33
- Pile of interlocking cubes
- Scrap paper

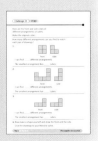

How many different arrangements can you find to match each pair of drawings?

Introducing the challenge

 Introduce the challenge to the whole class. Show only the front and side views of a shape you have made with interlocking cubes.

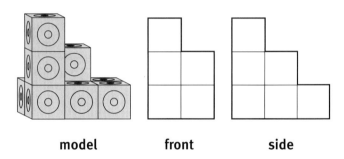

| model | front | side |

Ask the children to describe possible shapes. Make another shape, using their descriptions, encouraging them to be as precise as they can in their use of language. Compare the shape they have instructed you to make with the one you first made.

Look at the front and side views of my shape. Describe how I might have made this shape.

Look at these two shapes. How are they similar? How are they different?

Do they both have the same front view? What about the side view?

 Next, challenge individual children to come to the front and make a different shape with the same front and side views. Talk about what is possible and what is not possible.

Who can use these cubes to make another shape that has the same front and side views?

Has Masoud made another shape with the same front and side views?

How is his shape different from the other two shapes?

The challenge

 Provide each child with a copy of RS33, a pile of interlocking cubes and some scrap paper. Suggest that they draw the front and side of the models as they make them on scrap paper to check they look the same as the resource sheet. Encourage the children to be logical and systematic by asking questions such as:

If you turned it round just once like that, what will you see?

Which bit of the sides matches this bit of the front? How do you know?

How tall does the model have to be?

How wide does it have to be?

 Children make a drawing of their own model for a friend to make. They should keep their model hidden until it is time to compare the two models and the drawings.

Drawing out using and applying

 At the end of the challenge, discuss the difficulties and challenges of this problem with the children. Visualising can be hard for some and easy for others, and it is not always predictable who will find it hard or easy. Ask the children to suggest strategies that help, like looking for maximum height and width, rotating the model one face at a time, and so on.

Who was able to see an image of their model in their head? What did it look like?

Could you move the model about in your head? Did you have to physically make and move the model to see different arrangements?

Assessing using and applying

- Children can use the cubes and match the diagrams.
- Children can use the information from both views in a logical way to inform their model building.
- Children can extend their thinking from their first solution and use systematic and logical strategies to think of other solutions.

Supporting the challenge

- Children work in pairs to make the different models and find the different arrangements.
- Suggest the children make a model for a friend using no more than eight cubes.

Extending the challenge

- Children can make a set of similar problem activity cards for the class.
- Investigate problems with front, side and top views.

Squares in squares

Exploring number patterns in the context of shape

Using and applying

Reasoning

Explore patterns, properties and relationships and propose a general statement involving shapes; identify examples for which the statement is true or false

Communicating

Explain reasoning using diagrams and text; refine ways of recording using images and symbols

Maths content

Understanding shape

- Identify, visualise and describe properties of rectangles
- Use knowledge of properties to draw 2D shapes

Key vocabulary

square, possibility, generalisation

Resources

For each pair:
- RS34
- Scrap paper
- Ruler

For *Supporting the challenge*:
- 1 cm squared paper (see RS Extra)

Can Mother Hubbard give squares of fudge to any number of children?

Introducing the challenge

 Draw a large square on the board. Invite a child to the board to show how to divide the square into smaller squares, not necessarily all the same size, but with no pieces left over.

Include examples of different-sized squares.

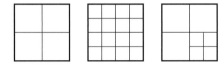

Can anyone divide the square into a number of smaller squares – for example, 12 smaller squares?

The challenge

 Provide each pair with a copy of RS34, some scrap paper and a ruler. Read through and discuss the challenge with the children and make sure they understand what they have to do. Encourage them to work together on the challenge.

To get smaller squares, children can divide any square within the large square into four smaller squares. This adds three squares each time (four smaller squares are created, but one larger one is lost). So starting with one, this number pattern builds up:

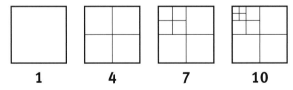

| 1 | 4 | 7 | 10 |

(They can also divide the square the same way, starting with nine or 25 smaller squares.)

Another possibility is to use the method suggested on RS34 to make a border of squares for any even number. For example, for 16 squares, make a border of 8 by 8 (the corner square is only counted once, so it balances with the large centre square to give an even number).

As the children work through the challenge, monitor their progress, asking questions similar to the following if they get stuck.

I wonder how Mother Hubbard could give squares of fudge to eight children? What about nine children?

How can you divide that square now to make some smaller squares? How many will that add? What if you did the same to another square? And another?

Can you invent a rule for making smaller squares? Where could you put the lines?

Drawing out using and applying

 Ask pairs of children to discuss their work. Focus not so much on the answers but on the methods the children used.

Let's collect all the different numbers of squares on the board. What is the smallest number we have? And the largest?

Can Mother Hubbard give squares of fudge to any number of children? What numbers of children can't she give squares of fudge to?

How can she give squares of fudge to 10 children? What about 11 children? Can you show the diagram?

Is this the only way of making 11 squares? How do you know?

Did you use trial and improvement? Did you have a plan?

Did you notice any patterns? What are they?

Can you make any generalisations?

Assessing using and applying
- Children can find solutions by trial and improvement.
- Children can use just one method to make some generalisations.
- Children can realise that by using different methods they can make all numbers above 7.

Supporting the challenge
- Provide children with 1 cm squared paper (see RS Extra) to draw out the squares.
- Help the children record their findings systematically. This should make it easier for them to see patterns and come to a generalisation.

Extending the challenge
- Investigate dividing rectangles into squares.
- Investigate dividing triangles into triangles.

Cross-stitch patterns

Transforming patterns using translation, reflection and rotation

Using and applying

Reasoning

Explore patterns, properties and relationships involving shapes

Communicating

Explain reasoning using diagrams and text; refine ways of recording using images and symbols

Maths content

Understanding shape

- Complete patterns with up to two lines of symmetry
- Draw the position of a shape after a reflection or translation

Key vocabulary

pattern, translation, reflection, rotation, clockwise, anticlockwise, right angle, quarter turn, half turn, symmetry, line of symmetry

Resources

- Examples of embroidery with symmetrical patterns or RS35 and RS36

For each child:

- RS37
- 0.5 cm squared paper (see RS Extra)
- Colouring pens
- Mirror
- Ruler

Can you make a simple cross-stitch design by translating, reflecting and rotating?

Introducing the challenge

 Discuss with the children about cross-stitch embroidery and how the patterns are made, often by taking a simple idea and working it into a more elaborate design.

Show the children some examples of embroidery with symmetrical patterns or use RS35 and RS36. You may wish to display the examples on an interactive whiteboard or print out an enlarged copy of the patterns. Ask the children to analyse the symmetry in the patterns, as well as any translations, reflections and rotations.

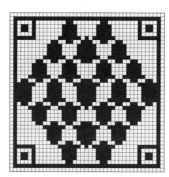

Look at these examples of cross-stitch patterns. Can you describe the pattern?

Look at this part of the pattern. What do you notice about the way it was made?

Can you show me an example of reflection/rotation/translation in this pattern?

The challenge

 Provide each child with a copy of RS37. Children start by analysing each of the patterns for their symmetry and looking at examples of transformations. They then design their own basic pattern, which should be quite simple, and repeat it in some way to make a border pattern.

If children have difficulty in coming up with a simple pattern, suggest to them one of the following:

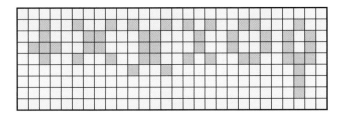

Mirrors can help with reflection, but less so with rotation. It may be useful to have the basic pattern on a separate smaller piece of paper so children can turn and flip at will.

After the children have made their simple pattern, ask questions that require them to think about how they are going to transform their pattern.

How are you going to rotate this shape? What will it look like after you rotate it a quarter of a turn in a clockwise direction? What about after half a turn?

Are you going to make one or more translations of this shape? How are you going to move it?

How will you reflect your shape? Along a vertical/horizontal line of symmetry? What about a diagonal line of symmetry?

Drawing out using and applying

 Have a display of the patterns and ask children to describe their pattern to the rest of the class. In particular, ask them to describe transformations they have used, using precise mathematical language.

Can you describe how you made your pattern?

Show us an example of symmetry in your pattern. What about where you translated your pattern? What about where you rotated it?

Assessing using and applying

- Children can analyse the cross-stitch patterns, recognise translation, reflection and rotation and use these words.
- Children can create their own patterns and transform them in a variety of ways.
- Children can describe their patterns and give 'rules' for the different forms of transformation so that another child could make the same border pattern from the description they have given.

Supporting the challenge

- Suggest the children make their border using translation.
- Suggest the children draw their basic pattern on a separate smaller piece of paper which they can turn and flip.

Extending the challenge

- Children make a border pattern that is continuous and involves reflection and rotation.
- Children write instructions for a pattern to be used as a frieze along a wall, so that their partner can make the pattern from the description and the basic design.

Nine squares

Drawing squares with different areas

Using and applying

Representing

Represent a problem by identifying and recording the information needed to solve it; find possible solutions and confirm them in the context of the problem

Reasoning

Explore patterns, properties and relationships, and propose a general statement involving shapes; identify examples for which the statement is true or false

Maths content

Understanding shape

- Identify, visualise and describe properties of rectangles; use knowledge of properties to draw 2D shapes

Measuring

- Calculate the area of a square

Key vocabulary

square, area, multiple, square units

Resources

For each child:

- Individual whiteboard and pen, RS38, Geoboards and elastic bands, grid paper: squared, isometric, triangular (see RS Extra)

> ## Can you make squares with areas that are equal to all the multiples of 9 to 90?

Introducing the challenge

 Invite a child to the board and ask them to draw a square with an area of 4 unit squares.

Invite another child to write up the multiples of 4 from 4 to 40. Ask the children to use their individual whiteboard and draw another square that has an area of a multiple of 4.

What is the area of your square?

Did anyone draw a square with the same area that looks different?

Who drew a square with an area of 20 square units?

The challenge

Provide the children with the resources. Introduce the challenge to the class. Ask the children to work in groups to create a set of squares that have areas equal to the multiples of 9. This needs to be a group challenge as children are likely to get discouraged easily when working individually.

Encourage the groups to keep a note of which squares they have found.

If groups find it difficult, you could direct their attention to the design decorating the top of the sheet.

What can you say about the area of the small square in relation to the large one? (It is half, so from a square of area 36, you can construct one of area 18.)

Drawing out using and applying

 Once the children have completed the challenge, bring the class back together again and ask individual groups to talk about their solutions and the way they went about finding the answers. Also encourage the children to talk about how their group worked together in completing the challenge.

Can you make squares with areas that are equal to all the multiples of 9 to 90?

Describe to us how you made the square with an area of 45 square units.

What strategies did you develop to help you solve the problem?

Can you apply these to other areas?

How did you work as a group?

How did people's role vary?

Assessing using and applying

- Children can work individually and only share their results.
- Children can work together but without planning how to deal with the challenge.
- Children can plan how they are going to go about the challenge and assign tasks to individuals or pairs.

Supporting the challenge

- Children use the Geoboards and elastic bands to make the squares.
- Children continue to investigate all the squares that can be made with areas that are multiples of 4.

Extending the challenge

- Can the same be done for other multiples: for example, 5?
- Investigate triangles on triangular grid paper (see RS Extra).

Egyptian circles

Working out the approximate area of a circle

Using and applying

Enquiring

Pursue an enquiry; present evidence by collecting, organising and interpreting information

Communicating

Explain reasoning using diagrams and text; refine ways of recording, using images

Maths content

Understanding shape

- Identify, visualise and describe properties of circles; use knowledge of properties to draw 2D shapes

Measuring

- Calculate the area of a circle

Key vocabulary

circle, octagon, area, square units, approximately

Resources

- NNS ITP: Polygon

For each child:

- RS39
- Plastic circles
- 1 cm squared paper (see RS Extra)
- Ruler

For *Supporting the challenge*:

- Overhead transparency of 1 cm squared paper

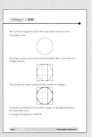

> **How good do you think the Egyptian method is for finding the approximate area of a circle?**

Introducing the challenge

 Using the NNS ITP: Polygon, display a square.

Ask the children to suggest different ways of finding its area.

How could you find the area of this square?

How else could you work it out?

Repeat for a rectangle and a triangle.

If children do not suggest the idea of calculating the area by counting the number of squares it occupies on squared paper, suggest this to the children and show them this on the ITP.

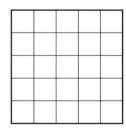

Once again, repeat for a rectangle and a triangle.

Making judgements about the reasonableness
of a solution and communicating this clearly

The challenge

Draw a circle on the board and pose the question:

How might you work out the area of a circle using a similar method?

 Discuss the various suggestions offered by the children.

Distribute RS39 and the other resources to the children and ask them to work individually to investigate the Egyptian method for finding the area of a circle.

The children may need some help in finding the area of the octagon effectively (it is seven-ninths of the area of the square). Rather than finding the areas of the small squares and the triangles separately, encourage the children to think of these in terms of their fraction of the whole.

 After enough time, children compare and discuss their results with a partner.

Drawing out using and applying

 Discuss the results of the investigation with the class.

How good do you think the Egyptian method is for finding the approximate area of a circle?

How close should an approximation be to be acceptable?

How can finding the area of an octagon help you find the area of a circle?

Assessing using and applying
- Children can present only the areas found.
- Children can judge the accuracy of the approximations by finding differences.
- Children can judge the accuracy of the approximations by comparing the ratio of the area of the circle with the approximation.

Supporting the challenge
- Provide the children with an overhead transparency (OHT) of 1 cm squared paper. The children place the OHT over the top of the plastic circles and help them construct a grid of nine squares to cover the circle
- Children draw around a plastic circle on squared paper. They count the number of complete squares entirely within the circle and also the number of complete squares that the circle just fits within.

Extending the challenge
- Use the method to find the approximations of larger circular objects.
- Investigate the history of ancient measuring systems.

Pave it

Investigating conservation of area

Using and applying

Solving problems

Solve problems involving shapes

Reasoning

Explore patterns, properties and relationships and propose a general statement involving shapes

Maths content

Understanding shape

- Identify, visualise and describe properties of rectangles; use knowledge of properties to draw 2D shapes

Measuring

- Calculate the area of a rectangle

Key vocabulary

area, square units, pattern, long, wide

Resources

For each pair:
- RS40 (optional)
- Dominoes or interlocking cubes joined in pairs
- 1 cm squared paper (see RS Extra)
- Ruler

How many different ways can you lay a path 2 units wide, using 10 paving stones?

Introducing the challenge

 Display six dominoes (or joined pairs of cubes) on the table. Ask the children to imagine that these are paving stones. The task is to pave a path that is exactly 2 units wide (a unit being half a domino or one cube).

Ask a child to lay out the dominoes to make a path which is 6 units long. For example:

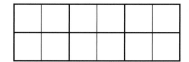

Can anyone find a different way to lay out the dominoes?

Can anyone think of another way?

How many different ways can we think of?

The challenge

 Provide each pair of children with a set of dominoes (or interlocking cubes joined in pairs) and a copy of RS40. Introduce the challenge to the children. Encourage the children to work together and to plan how they are going to tackle the challenge.

Before you start on the challenge, you need to think about how you are going to approach the problem. What might you do first? Why is this a good idea?

Remember, you have to lay out paths 2 units wide, using 10 paving stones. How might the paths you drew for five paving stones help?

 After a few minutes, stop the pairs and invite children to explain how they are going about the challenge.

If none of the children suggests it, present the idea that it might be easier to start with a simpler problem: two paving stones, three, four, and so on. Children look for a pattern and make a generalisation on the basis of that.

Making predictions related to area and shape, based on number patterns

 Provide children with 1 cm squared paper and a ruler and ask them to record their paths.

How many different ways did you find for laying five paving stones?

How many different ways are there for laying three or four paving stones?

Look at the different ways there are for three, four and five paving stones. How might these help you work out the different ways for laying six paving stones? What about seven?

Can you make a generalisation that will help you know how many ways there are for 10 paving stones?

Drawing out using and applying

 Gather together the results of the different pairs of children.

What is the number of possibilities for three paving stones? What about four paving stones? Six paving stones? 10 paving stones?

What patterns do you notice?

What generalisations did you make to help you find out how many possibilities there are for 10 paving stones?

Can anyone predict how many arrangements there will be for 20 paving stones? 35 paving stones? 100 paving stones?

Finally, ask different pairs of children to describe their recording methods and show them to the rest of the class.

Who used a table to record the number of different possibilities?

Who used another method of recording?

What was good about your method of recording?

Who would keep their record in a different way next time? Why? What might you do?

Assessing using and applying

- Children can work with 10 paving stones and randomly generate solutions, ordering their results afterwards.
- Children can work with 10 paving stones, but develop a systematic way of generating all the possible solutions.
- Children can realise for themselves that starting with simpler cases is an effective way into the challenge.

Supporting the challenge

- Children record their solutions for five paving stones on separate pieces of paper and organise these to check that they have found all possible solutions.
- Help children develop an effective method of recording their results.

Number of paving stones	2	3	4	5	6
Number of possibilities					

Extending the challenge

- What happens with a path 4 units wide?
- What happens with two different types of paving stones: single units and double units?

Cutting a strip

Estimating and measuring length

Using and applying

Solving problems

Solve problems involving length

Reasoning

Explore properties and relationships, and propose a general statement

Maths content

Measuring

- Read, choose, use and record standard metric units to estimate and measure length to a suitable degree of accuracy
- Draw and measure lines to the nearest millimetre

Key vocabulary

length, centimetre (cm), millimetre (mm)

Resources

For each child:
- Pencil and paper
- Ruler

For each pair:
- RS41
- Long strips of paper or rolls of till paper
- Scissors
- Tape measure showing centimetres and millimetres

For *Extending the challenge*:
- Furniture catalogues

How different are the three measurements? Why do you think this is?

Introducing the challenge

 Provide each child with pencil, paper and a ruler. Remind children what a hand span is.

span

 Ask each child to measure their hand span to the nearest millimetre and write it on the piece of paper. Tell the children to keep this a secret.

Children will probably measure their hand spans in different ways. Some may trace around their hand and measure this; others may place their hand span across the ruler. Allow the children to use whichever method they prefer.

 Pairs of children measure each other's hand span. Again, allow the children to use whichever method they prefer to measure their partner's hand span.

Ask each child to compare their two hand-span measurements: the measurement they did of their own hand and the one their partner did of their hand.

 Discuss with the children the differences in measurements.

Did anyone's partner get the exact same measurement as they did?

Who had a difference of plus or minus 1 millimetre? What about 2 millimetres?

Did anyone have a difference of a centimetre or more?

Why do you think we had differences in the measurements?

The challenge

 Provide each pair with a copy of RS41, strips of paper, scissors and a measuring tape. Introduce the challenge to the children. The children should not measure their height strips to start with, but match the length of the strip to their height by direct comparison.

Errors will inevitably creep in with each process: folding in half will cause one level of error, cutting another and measuring a third. The measuring will involve an element of rounding, either up or down for each piece, which will cause errors for the whole measurement.

Drawing out using and applying

 At the end of the challenge, discuss what the children have discovered. They may be surprised at the results. Compare the range of errors among all the children and ask for the children's ideas about what is an acceptable degree of error for what purpose. Consider how rounding up or down can cause errors if the measurement has to be magnified in some way.

Who got different measurements from their strips of paper?

What are the differences? Why do you think that is?

What degree of difference in measurements is acceptable? Is this always acceptable?

When is it important to be more accurate?

Assessing using and applying

- Children can accept and understand why errors occur in measurement.
- Children can see how magnifying measurements causes errors, both with rounding and with sampling.
- Children are beginning to understand that measurement is an approximation, and that an acceptable degree of error is inevitable.

Supporting the challenge

- Children use a tape measure rather than a ruler, as this is likely to provide greater accuracy when measuring lengths greater than 30 centimetres.
- Children cut strips of paper to the size of their spans. They use these to measure their height. How does this compare to their height when measured with a tape measure?

Extending the challenge

- Children look at packaged goods such as blocks in a box and consider the degree of accuracy for each block that is necessary when they are being made.
- Children look at measurements given in furniture catalogues. What degree of accuracy is usually acceptable?

How strong?

Exploring the relationship between force and weight

Using and applying

Representing

Represent a problem by identifying and recording the information needed to solve it; find possible solutions and confirm them in the context of the problem

Enquiring

Plan and pursue an enquiry; present evidence by collecting, organising and interpreting information

Maths content

Measuring

• Read, choose, use and record standard metric units to estimate and weigh to a suitable degree of accuracy

• Interpret a reading on a scale

Key vocabulary

weight, force, measure, heavy

Resources

For each pair:

• RS42 (optional)

• A collection of bags

• Rulers and tape measures

• A variety of weights and weighing instruments

How are you going to test the strength of your bag?

Introducing the challenge

 Show the children the collection of bags and discuss the different things that they are used for.

What else do we use bags for?

Why are there so many different types of bags?

Discuss how there are lots of different purposes for bags: for example, shopping bags, handbags, bags used in packaging goods such as fruit and vegetables, sugar and flour.

Discuss the relative strength of the bags.

The challenge

 Pose to the following question to the children:

How could you test the strength of these bags to see if they are strong enough for their purpose?

Briefly discuss the initial suggestions offered by the children. Write some of their ideas on the board.

Provide the children with a copy of RS42 and make sure that they have access to a collection of bags, rulers and tape measures and a variety of weights and weighing instruments. Ask the children to carefully choose a bag. Encourage the children to plan how they are going to test their bag.

How are you going to go about testing the strength of your bag?

Are there other ways you could go about doing it?

Which is the best way? Why?

How are you going to record the bags strength?

Drawing out using and applying

 Ask the children to explain to each other how they decided to test the strength of their bag.

Did the measures you used make it possible to compare different bags?

> Discuss which they think is the strongest bag.

Why do you think that this bag is the strongest?

How was this bag tested for strength?

Would we say that it is still the strongest if we tested it a different way?

Is there one method that could have been used for all the bags that would have made comparisons possible?

Assessing using and applying

- Children can test a bag to breaking point without considering its purpose.
- Children can realise that 'strength' could be interpreted in a variety of ways but only test one: for example, how long-lasting the material is or how water resistant it is.
- Children can coordinate the purpose of the bag with a range of strength tests.

Supporting the challenge

- Choose an appropriate bag for the children to work with.
- If pairs of children are having difficulty in deciding on an appropriate test for their bag, draw their attention to the list on the board and ask them to identify the one they feel is the most appropriate for their bag.

Extending the challenge

- Children design the strongest bag possible.
- Children investigate the strengths of various materials around the classroom.

Hands and feet

Comparing areas of irregular shapes and writing measurements to two decimal places

Using and applying

Reasoning

Explore properties and relationships, and propose a general statement involving numbers or shapes

Communicating

Explain reasoning using diagrams and text; refine ways of recording using images and symbols

Maths content

Measuring

- Measure and calculate area of irregular shapes
- Read, choose, use and record standard metric units to estimate and measure length to a suitable degree of accuracy

Key vocabulary

area, square centimetres, square decimetres

Resources

For each pair:
- RS43
- 1 cm squared paper (see RS Extra)
- Scissors

For *Supporting the challenge*:
- 1 square decimetre squares

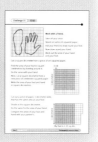

What is the area of your foot and hand in square decimetres?

Introducing the challenge

Discuss area with the children to make sure they understand what it is – there can often be misconceptions at this age.

Ask the children to look at various objects in the classroom and compare their areas. For example:

> The area of the whiteboard is greater than the area of one of the windows.

> The area of one of our desks is less than that of the classroom door.

> The cupboard door and the classroom door have a similar area.

The challenge

Provide each pair of children with a copy of RS43, 1 centimetre squared paper and a pair of scissors. Pose the following question to the children:

Which do you think has the greater area: your hand or your foot?

Children work in pairs to find out. As they work, discuss with the children methods for dealing with the part squares at the edge of their outlines and quick methods of calculating the area by breaking the area into rectangular sections.

You may want to stop the class from time to time to share methods different children are using.

The second part of the challenge is designed to show practically the relationship between units and subunits of measures. As the children work, ask:

How many square centimetres in a square decimetre?

What fraction of a square decimetre is a square centimetre?

How would you write 1 square centimetre as a decimal of 1 square decimetre?

Drawing out using and applying

 Share results with the whole class. Ask children to report on their results. Ask questions such as:

How much larger is your foot than your hand in square decimetres?

How much more than 1 square decimetre is your foot area? Your hand area?

How much less than 2 square decimetres?

How much more or less than your friend's?

Conclude by asking the children to discuss what they have learnt in this challenge about the standard metric units of length.

Assessing using and applying

- Children can work out the area of irregular shapes in a reasonable way.
- Children can show equivalent areas in regular shapes and know they are the same area.
- Children can record their work in decimal notation, using the notation to make comparisons, and explain their work.

Supporting the challenge

- Provide children with several 1 square decimetre squares to use.
- Mark 1 square decimetres into the corners of the centimetre squared paper for the children to use when measuring their hand and foot in square decimetres.

Extending the challenge

- Children find the average foot and hand area for the class and investigate whose is above and below average.
- Children make a square metre from square decimetres, use it to measure larger areas and record in decimal notation.

Round the world

Multiplying and dividing numbers related to time and distance

Using and applying

Enquiring

Plan and pursue an enquiry; present evidence by collecting, organising and interpreting information; suggest extensions to the enquiry

Communicating

Explain reasoning using diagrams, graphs and text; refine ways of recording using images and symbols

Maths content

Calculating

- Refine and use efficient written methods to multiply and divide

Measuring

- Read timetables and time using 24-hour clock notation; calculate time intervals

Key vocabulary

distance, time, travel, multiply, divide

Resources

For each pair:
- RS44 (optional)
- Calculator
- Information books
- Airline timetables
- Globe

Is it likely that the traveller went all the way round the world in 100 hours?

Introducing the challenge

 Pose the following statement to the children:

A traveller said: "I went all the way round the world to where I started, in 100 hours."

Briefly discuss children's initial thoughts.

What is your feeling about whether this is likely or not?

You could take a vote on the likelihood or have the children record 'yes' or 'no' against their names.

This is a problem where gut feelings can be useful, but children need to work it out to verify their instincts. Gather ideas from the children about how they might begin to work it out, then leave them in pairs to do so.

What is the first thing you need to think about?

How will you go about finding this out?

What things will you need to make sure that you can answer this problem accurately?

The challenge

 The children need to draw on various bits of knowledge and understanding: for example, how long plane journeys take, the distance round the earth at different latitudes, where different places are on the globe, how long planes can travel before they need to refuel. They also need to think about how to assemble all this information.

Children need to decide on what calculations to do. A well-justified estimate will work for this problem.

Draw the children together from time to time to discuss and clarify aspects of these points. With the class, establish a definition of 'round the world'.

Drawing out using and applying

 Because there is no way of guessing the answer to this quickly, the children need to justify their answers and double-check their working out. The children should present their work to the group. Encourage the group to cross-examine the evidence.

Is it likely that the traveller went all the way round the world in 100 hours?

Why do you say that?

How did you work that out?

What evidence have you got to support that statement?

Who disagrees with this? Why?

What sources of information helped you arrive at your decision?

What else would have helped?

How did you work out how long it would take to travel round the world?

Did anyone work it out using a different method?

Who would you like to ask Neela a question?

Assessing using and applying

- Children can decide what information they need and talk to each other about it.
- Children can do the calculations and present their results in a clear and understandable form.
- Children can justify their conclusions to the class, answering challenging questions in a convincing way.

Supporting the challenge

- Some children may be daunted by the challenge and not know where to start. Assist them in accessing the problem by directing them to appropriate resources and asking questions that will help them get started.
- Help the children with calculating time intervals.

Extending the challenge

- Children look at the story of 'Around the world in 80 days' by Jules Verne and investigate the timings of the journey, considering the International Date Line.
- Children plan journeys to places they would like to visit, with a set time limit: for example, 8 hours or 32 hours.

A fair six?

Carrying out simple probability experiments

Using and applying

Reasoning

Explore patterns, properties and relationships and propose a general statement involving numbers; identify examples for which the statement is true or false

Communicating

Explain reasoning using diagrams and text; refine ways of recording using images and symbols

Maths content

Handling data

- Describe the occurrence of familiar events using the language of likelihood
- Answer a set of related questions by collecting, selecting and organising relevant data; draw conclusions and identify further questions to ask
- Construct frequency tables and bar graphs to represent the frequencies of events

Key vocabulary

fair, unfair, probability, chance, likelihood

Resources

For each pair:

- RS45 (optional), 1–6 dice, 1 cm squared paper (see RS Extra), ruler

Is a 6 a difficult number to roll on a 1–6 dice?

Introducing the challenge

 We recommend that the children complete this challenge before attempting Challenge 34: How likely?

Discuss games the children know, such as board and card games, and the rules of those games. There are general conventions that apply to all games, such as taking turns in order, and specific rules that apply to certain games, such as having another go if you roll a 6.

The children should consider how these rules contribute to the game's fairness. Clearly, in a game of chance, as long as everyone plays with the same rules, it feels as if everyone has the same chance of winning.

When playing board or card games, what are some of the basic rules?

Why do we have these rules?

How would the game be different if we didn't have these rules?

The challenge

 Arrange the children into pairs and provide each pair with the necessary resources. Discuss the idea of 'special' numbers with the children.

Do you feel that there is something difficult about rolling a 6 on a 1–6 dice?

Ask the children to suggest ways of testing this. They could roll the dice many times and record what was rolled; several pairs of children can do this and combine the results. Do they have 'lucky' numbers that come up for them? If a child says that 3 is her lucky number because she 'always gets 3', then she must test that.

Investigating and making general statements

 Children work in pairs and find systematic ways of rolling the dice and recording the number, and they need to do this over 100 times before the results are worth looking at. You may need to discuss this with the children if they stop at, for example, 20 rolls, only having rolled one 6. You may need to collect enough evidence to draw the conclusion that there is a 1-in-6 chance of rolling a 6 (or any other number) on a 1–6 dice.

The children should be able to invent visual ways of recording their findings, such as graphs on squared paper.

Drawing out using and applying

 Pairs of children share their ideas with the rest of the class. Encourage them to produce evidence to support or refute the premise that a 6 is more difficult to roll than any other number on a 1–6 dice.

Is a 6 a difficult number to roll on a 1–6 dice?

Is it any more or less difficult than any other number?

What evidence do you have to support your claim?

Does anyone disagree with this? Why?

Is the likelihood of rolling a 6 on a 0–9 dice the same as for that on a 1–6 dice? What makes you say that?

Assessing using and applying

- Children can see ways of testing the statement that a 6 is more difficult to roll.
- Children understand that the more tests they do, the 'better' the result will be.
- Children can make statements as a result of their experiments, such as "A 6 isn't any more difficult to roll than a 3: all the numbers have the same chance of coming up."

Supporting the challenge

- Help the children draw a simple table to organise their results:

Dice number	Frequency
1	
2	
3	
4	
5	
6	

- Help the children label the axis for a bar chart displaying their results:

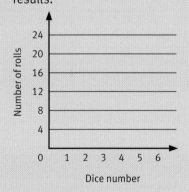

Extending the challenge

- Children investigate events with a probability of 1 in 2.
- Children investigate the probability of getting 12 when adding the score of two 1–6 dice.

How likely?

Discussing events and considering their likelihood

Using and applying

Reasoning

Explore relationships and propose a general statement; identify examples for which the statement is true or false

Communicating

Explain reasoning using diagram and text

Maths content

Handling data

- Describe the occurrence of familiar events using the language of likelihood
- Answer a set of related questions by collecting, selecting and organising relevant data

Key vocabulary

likelihood, certain, likely, equally likely, unlikely, impossible

Resources

- NNS ITP: Number spinners

For each pair or group:

- RS46
- Pencil and paper
- Scissors

What is the likelihood that it will rain today?

Introducing the challenge

 We recommend that the children have completed Challenge 33: A fair six? before attempting this challenge.

Display the square spinner on the NNS ITP: Number spinners and number the spinner 1, 2, 3 and 4. Discuss with the children the likelihood of spinning the following:

- the number 2
- an even number
- a number less than 5
- a number greater than 5

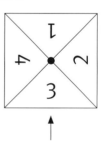

Change the numbers on the square spinner to 1, 2, 1, 3, and ask questions similar to those above.

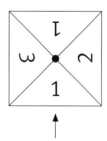

Repeat the above for another spinner shape. For example:

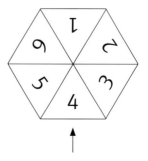

As the children discuss the likelihood of spinning particular numbers, write up the following key words they use: certain, likely, equally likely, unlikely, impossible.

Children suggest statements that are certain, likely, equally likely, unlikely, impossible and write at least one example of each on the board.

- **Certain:** *If I drop this book, it will fall downwards.*
- **Likely:** *I will watch some television tonight.*
- **Equally likely:** *If I toss this coin, it will land heads up.*
- **Unlikely:** *It will snow tomorrow.*
- **Impossible:** *If I snap my fingers, you will turn into a frog.*

The challenge

 Arrange the children into pairs or groups of three and provide them with a copy of RS46, pencil and paper and scissors. The children work as a group to invent their own statements and match them to each of the five categories.

They then swap statements with another pair or a group of three and sort each other's statements into the five categories.

The groups then swap back, look at the way their own statements have been sorted and justify any sorting they disagree with. As the children work, ask them to justify the decisions they make.

What is the likelihood of this statement? Why do you think that?

Did both groups agree the likelihood of all of your statements? Which ones did they disagree with?

Drawing out using and applying

 At the end of the challenge, discuss with the children any unresolved disagreements about statements. If some categories are under-represented, such as 'certain' or 'equally likely', ask for more statements and work some out as a class.

Who thinks that this statement is likely? Does anyone disagree? Why?

Can someone tell me another unlikely statement?

Assessing using and applying

- Children can give examples of each kind of statement, including 'equally likely'.
- Children can show that they understand the statements as they are using them and talking about them.
- Children can explain the logic behind their thinking: for example, "I know it's equally likely my aunt will have a girl or a boy, because with having babies, it could just as well be either."

Supporting the challenge

- Provide the children with a list of some statements and ask them to sort them into the five categories. Then ask them to add one or two more statements for each category.
- Children write about the likelihood of rolling particular numbers on different dice.

Extending the challenge

- Investigate the risks involved in doing certain things such as playing with matches, cutting towards you with an open knife, crossing the road, sitting next to someone who has a cold, and so on. Collect and analyse national statistics on similar issues.
- Make a collection of phrases that are related to probability, such as: '50 – 50'; 'It's six of one and half a dozen of the other'; 'Once in a blue moon'; 'Over my dead body'; 'It never rains but it pours' ...

Words in a newspaper

Gathering data and displaying it appropriately

Using and applying

Representing

Represent a problem by identifying and recording the information needed to solve it; find possible solutions and confirm them in the context of the problem

Enquiring

Plan and pursue an enquiry; present evidence by collecting, organising and interpreting information; suggest extensions to the enquiry

Maths content

Handling data

- Answer a set of related questions by collecting, selecting and organising relevant data; draw conclusions and identify further questions to ask
- Construct frequency tables and graphs to represent the frequencies of events

Key vocabulary

data, information, table, collect, organise, present, analyse, interpret, estimation

Resources

For each group:

- RS47 (optional)
- Newspaper
- Calculator

How many words does a newspaper contain?

Introducing the challenge

 Pose the following question to the children:

How many words do you write on one page of an exercise book?

Discuss the question with the children and the fact that the number of words would vary depending on the subject. For example, there are probably few words in a maths exercise book, but probably a lot of words in a writing book. Also, the number of words per page depends on the size of the writing.

Write some of the children's estimates on the board and show the children some exercise books to help them in their estimations.

Pose the following question to the children:

Can you estimate how many words you write in an exercise book?

Once again, discuss the question with the children and the fact that the number of words would vary depending on the factors discussed previously, as well as on the number of pages in the book.

The challenge

Hold up and flick through a copy of a newspaper.

Pose the following question to the children:

How many words do you think are printed in a newspaper? Hundreds? Thousands? Millions?

Children note down individually how many words they think there are. Collect together some of the estimates. Discuss the figures.

How big is the spread between the largest and the smallest estimates?

 Arrange the children into groups. Introduce the challenge, using RS47 if appropriate. Give the children a few minutes to brainstorm their ideas on how they could make a good estimate of the number of words in the paper.

Bring the class back together and invite children to share their ideas.

Which methods do you think will give the most accurate results?

How are you going to cope with different print sizes?

Are you going to include advertising or simple news stories?

How might you go about recording the information that you find out?

What are the different ways you could present your findings?

Which might be the most appropriate? Why?

 Children go back into their groups and plan how they will solve the problem. Explain that they can have a copy of a newspaper when they have shown you their plans.

Drawing out using and applying

 Gather the groups together to share their findings. Discuss with the class the way different groups recorded their findings and which were the easiest to interpret and why.

How many words does a newspaper contain?

If everyone was working on different copies of the same paper, ask:

Did any group come up with a radically different number?

If everyone was working on different newspapers, ask:

How big a difference does there seem to be between papers?

How did you organise yourselves to work?

Did any group organise themselves differently?

Which method of organisation do you think is the most efficient? Why?

How did you present your results? Did this make it easy to read and interpret?

How could you have presented your findings differently? Would this have been better? Why?

Assessing using and applying
- Children do not organise themselves effectively and duplicate work.
- Someone in the group takes a leadership role and organises the group.
- Children can discuss various strategies and agree on the best approach.

Supporting the challenge
- Arrange the children into mixed-ability groups.
- Help the children organise and present their results.

Extending the challenge
- Based on their estimates, how long would it take one person to write out the newspaper?
- Children investigate the occurrence of letters: which is the most frequent?

Channel hopping

Planning, collecting, organising and interpreting data, including finding averages

Using and applying

Enquiring

Plan and pursue an enquiry; present evidence by collecting, organising and interpreting information; suggest extensions to the enquiry

Communicating

Explain reasoning using diagrams, graphs and text; refine ways of recording using images and symbols

Maths content

Handling data

- Answer a set of related questions by collecting, selecting and organising relevant data; draw conclusions and identify further questions to ask

- Construct frequency tables and graphs to represent the frequencies of events; find and interpret the mode of a set of data

Key vocabulary

data, information, collect, organise, present, analyse, interpret, table, graph

Resources

- For each group:
- RS48 (optional), pencil and paper, 1 cm squared paper (see RS Extra), ruler

For *Extending the challenge*:

- TV guides

How often do you and your family 'channel hop'?

Introducing the challenge

 Discuss with the children the role that television plays in their lives. Brainstorm the different things that the children could investigate about television. Write some of these suggestions on the board.

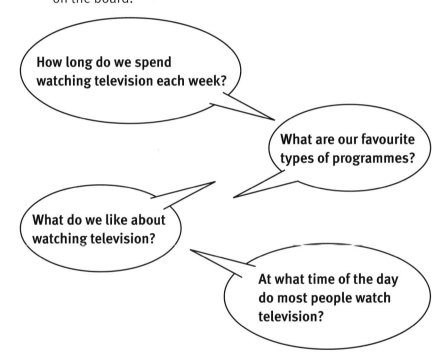

How long do we spend watching television each week?

What are our favourite types of programmes?

What do we like about watching television?

At what time of the day do most people watch television?

Briefly discuss one or two of the questions in more detail, focusing on how the children would go about collecting, organising and presenting data to answer the line of enquiry.

Also, briefly discuss with the children the term 'average', and how they could find the average of one of the previously discussed topics.

The challenge

 Discuss the idea of the survey with the children, using RS48 if appropriate. Ask the children for their ideas about why this information might be useful for programme makers and advertisers. Brainstorm with the children ideas for collecting the information, and suggestions for a useful data-collection sheet.

Explain to the children that they are to work in groups of four or five to plan and carry out the survey themselves. Talk through the instructions on RS48, which record each group's planning of the activity. Set an agreed day by which the data will be collected.

 The first part of the work is planning what data to collect, how to collect it and what to do next. The planning is an important part of this challenge, and children may need help in structuring it. Whenever necessary, bring the whole class back together for a discussion of the organisation of the challenge. You may want to discuss further the structure of a data-collection sheet, as this is crucial to the success of the challenge. What the children decide to do with the data and how they represent it depends on their previous experience with data handling.

When the children have collected the data, they work within their groups to gather the data together onto one information sheet and calculate the average. You may need to help some groups at this stage. Each group prepares a formal presentation to the whole class, with a written account of their procedures, graphs, diagrams or charts, where appropriate, and a prepared oral delivery.

Drawing out using and applying

 Gather the groups together to share their findings. Discuss with the class the way different groups recorded their findings and which were the easiest to interpret and why.

On average, how often do we 'channel hop'?

Did any group get a very different answer? Why do you think this was?

How did you organise yourselves to work?

Who organised themselves differently?

Which method of organisation do you think is the most efficient? Why?

How did you present your results? Did this make it easy to read and interpret?

How could you have presented your findings differently? Would this have been better? Why?

Assessing using and applying

- Children can plan together to decide what information they need, how to collect it and what to do with the data.
- Children can work together to discuss the progress of the work and to continue with the agreed objectives.
- Children can present their work in writing, pictorially and orally and explain it clearly.

Supporting the challenge

- Help the children plan their investigation.
- Help the children create their data-collection sheet and finding the average.

Extending the challenge

- A related survey is to find out how long viewers take on average to decide whether a programme they have just switched to is worth watching.
- Children look at planned televisions viewing instead. They plan a week's television viewing and present it on a timetable. They also include a plan for videoing programmes they cannot watch and plan in viewing these, too.